UNDERSTANDING SCIATICA

TOM JESSON

with ANNINA SCHMID

Illustrated by PETER JESSON

And TOM JESSON

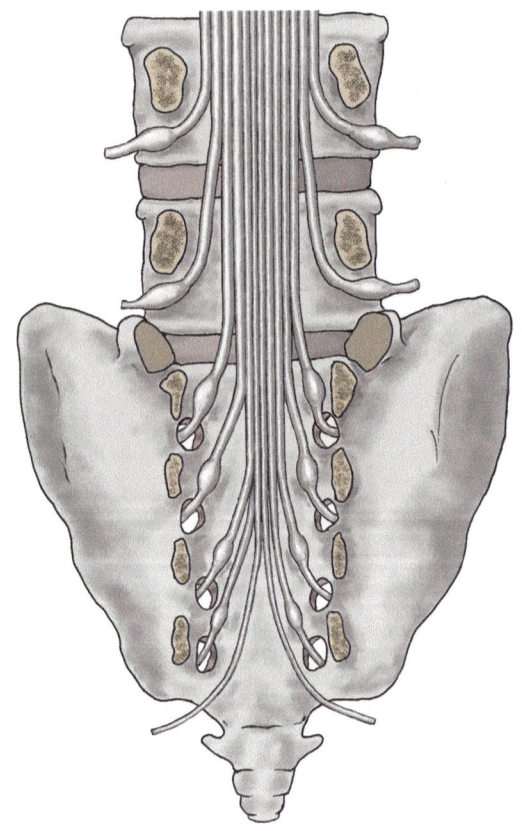

PRAISE FOR UNDERSTANDING SCIATICA

'Absolute gold... There's not many anatomy/physiology books that one can't put down' - June, specialist physiotherapist in spinal triage.

'I was expecting a good read, but what you get is even better! Absolutely brilliant stuff and a MUST read for any clinician who works with sciatica' - Michael, physiotherapist.

'Simply a good book as well as a good medical book. So much contemporary content, blended with a personal, readable format like the classic medical texts of the past.' - Greig, Sports & orthopaedics doctor.

'The book has made a massive difference to my practice and in explaining things to my patients.' - Philip, spinal advanced practice practitioner.

'Highly recommend this book. In addition to being informative and engaging it provides sensible, well-reasoned insight, useful for patients and clinicians alike' - Michael, US military PT.

'The sciatica Bible' - Luke, physiotherapist.

CONTENTS

FOREWORD BY ANNINA SCHMID

Sciatica is a complex condition. This becomes immediately apparent as I listen to the stories of people with sciatica in my research and clinical practice. Some report tingling, burning pain and electric shocks in their leg, while others experience a numb foot. Some find sitting impossible while for others this position is the only one that relieves symptoms. Some people have muscle weakness and sensory loss, while others experience cramps or strange feelings such as water trickling down their leg. Some people are highly functioning whereas others can hardly get out of bed. Some people recover spontaneously while some develop persistent pain with repeated flare ups. Some people benefit from physiotherapy or anti-inflammatory medication, while others' pain remains resistant to such treatment. Yet we summarise all this heterogeneity under the same label of 'sciatica'. It is therefore not surprising that it remains a challenge to make sense of 'sciatica' not only for people who experience this condition, but also for clinicians.

In research, we have made substantial advances in our understanding of sciatica and nerve injuries in the past years. However, as researchers we are sometimes so immersed in our complex theories, hypotheses and data, that we struggle or lack time to communicate our findings in an easily digestible way. After all though, science is useless if our discoveries and their clinical relevance are not translated and communicated to clinical settings.

In the 'sciatica world', we are truly lucky to have Tom Jesson. This book is a stellar example of how understanding the basic scientific principles of neuroanatomy, biomechanics, neurophysiology, biology and pain neurosciences can help us understand 'sciatica' and its clinical nuances and distinct presentations. This book thereby elegantly juxtaposes complex scientific material and clinical pearls. It is a joy to see complicated research translated into clinically relevant and digestible concepts. Importantly, Tom manages this without compromising on accuracy. His enquiring, critical mind and outstanding ability to extract the essence from scientific texts are deeply reflected in this book.

It was an absolute pleasure to have contributed a small part to the refinement of this book. I am convinced it will lead to many 'aha' moments for its readers-as it has for me. Ultimately, this book and the knowledge it consolidates will help us be better clinicians, something we are all striving towards. I hope you will enjoy reading it and discovering that cleverly marrying science with clinics is highly exciting!

'*The job of a clinician is to shut up, listen, care and know something.*'

— ANON

(This book will help you with that last one!)

PART I: INTRODUCTION

1

REFERRED PAIN, RADICULAR PAIN, RADICULOPATHY... AND SCIATICA

'This is the history of medicine: giving a thing a name and then everyone thinks they know all about it...'

— PETER NATHAN, 1977

In 1994, the radiologist Pierre Milette wrote to the journal Radiology on the issue of referred pain, radicular pain and radiculopathy to ask simply, 'What are we really talking about?' (1).

According to Milette not only were clinicians confused about the terminology, but so were academics. 'If we seek to improve our understanding,' Milette wrote, 'it is mandatory to address this fundamental issue.'

Let's do that now.

Referred pain

Referred pain is felt in a part of the body remote to that of the original injury. The original definition from the International Association for the Study of Pain is 'pain perceived as arising or occurring in a region of the body innervated by nerves or branches of nerves other than those that innervate the actual source of pain' (2).

Why does this happen? The standard theory is that when danger messages (nociceptive signals) from an injury arrive in the spinal cord, they are mixed up with normal messages from other parts of the body. Those danger messages and normal messages are passed up the spinal cord to the brain together. The brain is unable to tell the two apart and creates a pain experience for both.

That's the standard theory, and it's perhaps a little over-simplified, but it gets at the basic idea of referred pain: the brain is 'confused' about the exact location of the problem.

There are two types of referred pain. First, visceral referred pain is caused by danger messages from internal organs like lungs, intestines, and kidneys. For example, danger messages from the spleen can be felt in the shoulder, and danger messages from the heart can be felt in the left arm.

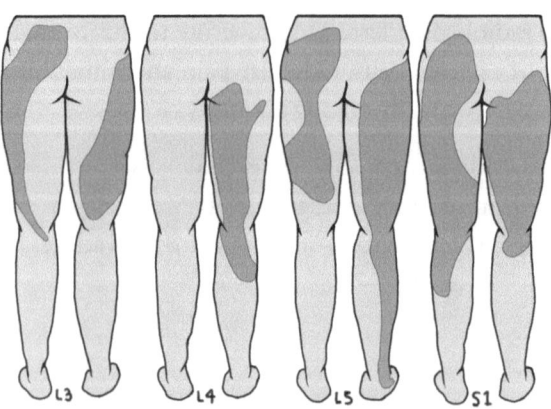

Patterns of somatic referred pain. Adapted from a 1939 experiment, in this picture you can see some examples of patterns of referred pain originating from noxious stimulation of the lumbar interspinal ligaments at different segmental levels (3).

4

Second, somatic referred pain is caused by danger messages from somatic tissues like bones, cartilage and muscles. For example, danger messages from an intervertebral disc or a facet joint can be felt in the buttock and down the leg.

Referred pain is dull, aching, gnawing, often deep and difficult to localise.

Radicular pain

Radicular pain is a kind of nerve pain. It's caused by action potentials that emanate from the nerve root and/or the dorsal root ganglion (4). This, of course, is not where action potentials are supposed to come from. Ordinarily, they should start in nerve endings in target tissues such as skin, bone, muscle and so on.

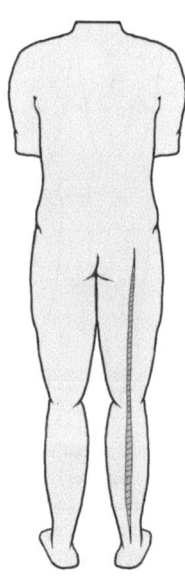

Action potentials that emanate from within a nerve or its dorsal root ganglion are called 'ectopic impulses' (ectopic means 'in the wrong place').

In addition to these ectopic impulses, part of the clinical picture of radicular pain is likely also caused by a more generalised neuronal hyperexcitability. In response to injury at the nerve root, a neuron can amp up its defence strategy so that even normal, non-ectopic impulses are sparked more readily in response to stimuli.

One way of representing the 'classic' radicular pain pattern. Other representations are more dermatomal. Adapted from Bogduk (4).

How does radicular pain feel? Classic radicular pain *roughly* tracks the territory of the affected nerve root. The pain is sharp, shooting, stabbing and usually severe. Often, the pain is accompanied by a dull

5

or burning background ache, as well as 'nervey' sensations like pins and needles and tingling.

Radiculopathy

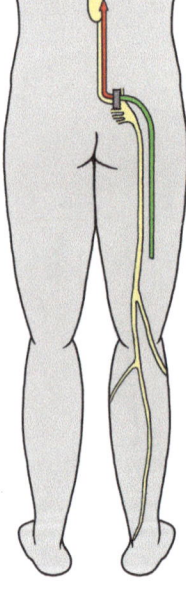

Painful radiculopathy. Green line = normal impulses from the nerve tips: they are blocked or slowed at the injured nerve root, and will not (all) get to the brain. That's a radiculopathy. Red line = aberrant impulses emanating from the injured nerve root, and these *do* get to the brain, causing pain. That's radicular pain

Radiculopathy is another nerve problem. However, it is *not* a pain condition. Instead, the term describes *loss* of nerve function. 'Loss of nerve function' means that fewer action potentials are conducted up and down the injured nerve because of an injury to the nerve root or dorsal root ganglion. The nerve isn't doing its job.

A loss of nerve function is a pretty common everyday experience. If you sit too long and your leg goes numb, that's a loss of nerve function. A radiculopathy is not too different, although of course it involves the nerve root in the spine rather than the nerve trunk in the periphery, and often involves more lasting damage to the nerve too.

Radiculopathy manifests as a dulled or absent reflex response, a loss of sensation to different sensory stimuli (e.g., touch, sharp prick, warm/cold), and/or a loss of muscle strength.

The clinical picture is less clear cut

Of course, referred pain, radicular pain and radiculopathy can all occur together; all three can overlap

First, although radicular pain and radiculopathy can occur separately, they often co-exist as a 'painful radiculopathy'. This makes sense, of course - they both involve a problem with a nerve root, so it's not surprising that such a problem can cause both pain and loss of function. Sometimes the radicular pain is serious and the radiculopathy is mild, and sometimes it's the other way round - and everything in the middle.

And referred pain can also co-exist with radicular pain and radiculopathy. If you think about it, this makes sense. Imagine a big disc herniation that injures a nerve root and causes radicular pain. That disc herniation and all the associated inflammation might easily also trigger nociceptive signals from the disc and the surrounding tissues. That nociception might cause referred pain in the buttock or down the leg.

So, where there is radicular pain there can also be some somatic referred pain.

These mixed pain presentations partly explain why so many cases of radicular pain do not look like they are 'supposed' to look: it's often not only radicular pain, but radicular *and referred* pain.

Even aside from the presence of referred pain however, radicular pain itself can deviate from the 'classic' picture of a band of pain that tracks the territory of the affected nerve root, i.e. a dermatome of pain. In fact, numerous studies show that it's near-impossible to tell from the pain pattern alone what nerve root is causing pain (5–7); radicular pain doesn't obey the textbooks' dermatomes. On top of that, radicular pain can also expand beyond the expected band of pain to a wider territory (8).

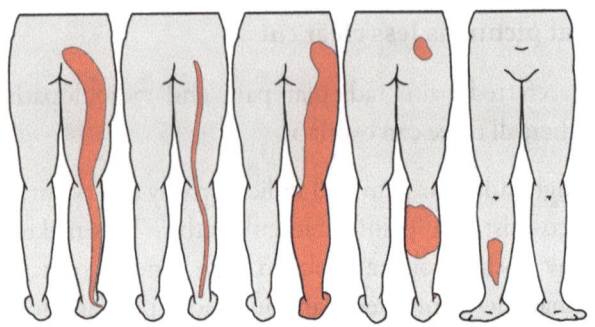

Varieties of nerve root pain. From left to right: Classic band-
like, dermatomal pain, often shown in textbooks; classic
thinner band of pain, as described by Bogduk (4); extra-
territorial spread; 'jumping' pain; occult patches of pain.

Sometimes, nerve root pain doesn't form a continuous line at all, but leaps from patch to patch, for example from the buttock to the shin to the big toe; or even just show up in one patch - *just* the buttock, or *just* the shin, or *just* the big toe (9). We don't want to overstate this variety; the classic picture is the classic picture for a reason. But it's important to know that radicular pain, like most pain, can present atypically.

Sciatica

> *'Amongst painful diseases, sciatica occupies a foremost place by reason of its prevalence, its production by a great variety of conditions, the great disablement it may produce, and its tendency to relapse; all of which have long ago led to its recognition as one of the great scourges of humanity'*
>
> — - VITTORIO PUTTI, 1927 (10)

The word 'sciatica' is less a diagnosis and more of a vague gesture - 'there's pain in the back of the leg... for some reason'. It doesn't really have any official definition and different people mean different things by it. It's a throwback to a time when we had much less medical knowledge. The eminent spinal surgeon Jeremy Fairbank even called sciatica 'an archaic term' (11).

One problem with 'sciatica' is that it means different things to different people. Of course, the same could be said for 'referred pain', 'radicular pain' and 'radiculopathy', which, as Milette complained, are often used carelessly. But for those words there is at least an official definition to refer to. Not so for sciatica. To some people, it means radicular pain, to others it means any pain that comes from the spine but is felt in the leg, including referred pain, and to still others it means any pain in the back of the leg that seems to have something to do with a nerve (12).

That said, sciatica is a recognisable word for laypeople, which is important. And it's a useful word for clinicians who want to refer to everything we've looked at above without having to say 'referred pain or radicular pain or radiculopathy or some mixture of the three'! It's kind of a catch-all term, in that respect. We use it all the time - in the title of this book, for example! So, whereas the scientific community discourages the use of the term 'sciatica' (13), its widespread use by laypeople and its usefulness as a colloquial catch-all means it is likely to stick around.

That's enough place-setting! Let's get started.

Key points on referred pain, radicular pain and radiculopathy:

- Referred pain is when pain from tissues like muscles, joints and discs is felt in the wrong place. It is usually a diffuse ache.

- Radicular pain is when pain from the nerve root in the spine is felt roughly in the territory of that root. In the case of lumbar radicular pain, that's down the leg. Radicular pain is usually sharp and severe.
- Radiculopathy is when an injury to the nerve root stops it from conducting impulses to and from the brain. This makes muscles weaker and sensation duller.
- Because all of these things can exist together, in different amounts, the clinical picture is often far from clear! Additionally, radicular pain itself has a varied presentation, not always appearing in the expected dermatome.
- Sciatica is an old-fashioned term without any specific meaning. Despite this, it is an easy way to refer to pain down the back of the leg that seems to be related to a nerve.

2

AN ANATOMY TOUR

Let's start with a tour.

On this tour, we will follow the primary sensory neurons. We will start at the top, where each primary sensory neuron synapses in the spinal cord in the upper lumbar spine, and many of them are bundled together as rootlets. Then, we'll follow these neurons downward through the spine as they bundle together again into nerve roots, then jumble together into the spinal nerve, and finally exit the spine through the intervertebral foramen. Finally, we'll continue to follow them down as, now bundled together as peripheral nerves, they continue their journey to the tips of the toes and other tissues.

Rootlets become roots; many roots make up the cauda equina.

In infants, the end of the spinal cord is at about the L3 vertebral level. But as we age, it is outgrown by the rest of the body and by the time we are in our teens, the end of the spinal cord is further up the spine, at L1 or L2. The cord tapers to terminate at the conus medullaris.

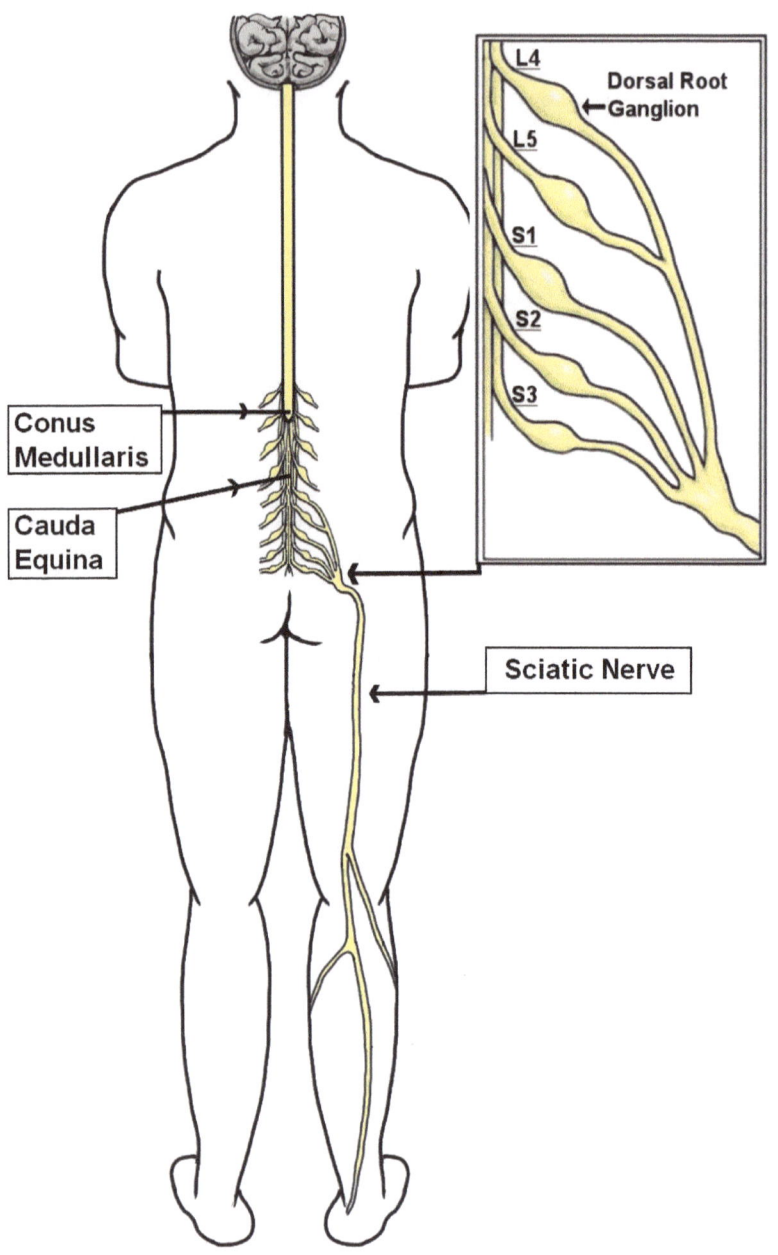

Conus
Medullaris

Cauda
Equina

L4

Dorsal Root
←Ganglion

L5

S1

S2

S3

Sciatic Nerve

The nerve roots enter and exit the spinal cord not as fully formed roots, but as rootlets. There are dorsal (posterior) rootlets, which are made up of sensory neurons carrying impulses from the body and its environment up to the spinal cord. And there are ventral (anterior) rootlets, which are made up of motor neurons carrying impulses from the spinal cord down to the muscles.

After the rootlets bud off from the spinal cord, they form sub-bundles, and then form bundles once more to become the nerve roots that make up the hanging tail of the cauda equina. When they bundle, dorsal rootlets stick together and ventral rootlets stick together, making separate dorsal sensory roots and ventral motor roots. Dorsal roots are thicker than ventral roots because there are more sensory than motor neurons in peripheral nerves.

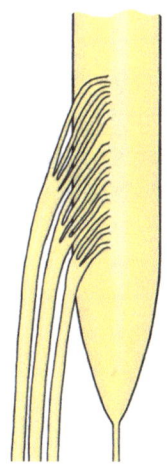

Schematic diagram showing how the cord tapers into the conus medullaris. The L4, L5 and S1 dorsal rootlets bud off the spinal cord and bundle to become nerve roots.

The lumbar spinal cord is small, about the size of a little finger, and the roots are smaller still: between two and four millimetres in diameter in the lumbar spine. When an anatomist is laying the roots out or holding them, they seem like tangled spaghetti, hanging off a fork. In the body, where they are guided and held by ligaments and connective tissue, they have a more orderly and linear appearance, like uncooked spaghetti still in the packet (although of course they are not hard but soft and pliable, with the consistency of rubber).

The cauda equina is inside the dural sac.

Everything we have seen so far is taking place inside the protective dural sac (sometimes called the thecal sac). The dural sac envelops the spinal cord in the cervical and thoracic spine and, after the spinal cord terminates in the upper lumbar spine, it descends into the lumbar spine to protect the nerve roots as they make up the cauda equina.

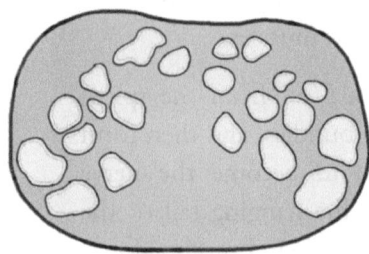

Cross section of the lumbar and sacral nerve roots in the dural sac, adapted from Cohen and colleagues (14)

The dural sac has two layers. The outer layer is made up of the tough dura mater ('dura' comes from the same root word as 'durable'). The inner layer is the thin, transparent arachnoid membrane.

Beneath the arachnoid membrane, pumping back and forth on its slow course around the brain and spine and bathing and nourishing the cauda equina, is the cerebrospinal fluid (CSF). The pressure of the CSF helps to balloon up the arachnoid membrane and dura mater into a plump sac shape (after it's been dissected from the body and it's lost its CSF, the dural sac is more of a sad, flat sleeve). As is evident from pictures of the dural sac in cross-section, the roots have quite a lot of space and shift around as we move.

Between the CSF and the nerve roots is the innermost layer of protection, the delicate pia mater. The pia mater, which covers the brain in the skull and the spinal cord in the cervical and thoracic spines, surrounds the individual nerve roots in the lumbar spine. If you were to remove the pia mater from a nerve root and roll the root in your fingers, it would de-bundle like loose twine into the separate rootlets that we originally saw budding off from the spinal cord.

If it was not already clear, these three layers - the dura mater, the arachnoid membrane and the pia mater - are the same contiguous membranes that cover the brain, too.

At each spinal level, roots change direction and head for the intervertebral foramen.

At each spinal level on the cauda equina's downward course through the dural sac, one dorsal and one ventral root per side will pair up and branch off together. They will go forth and innervate their particular patches of the low back and leg. As they branch off, the pair of roots take with them a portion of the dural sac, which will now form the nerve root sleeve. The sleeve binds the two roots more tightly than did the dural sac, holding them on course. As the space inside the sleeve is continuous with the space inside the dural sac, there is CSF in here too, nourishing the nerve roots.

In the short part of their course after they leave/enter the dural sac and before they leave/enter the spinal column completely, the nerve roots are at their most vulnerable. Although these 'extra-dural, intra-spinal' roots still have the protection of the dura mater and the arachnoid membrane (now in the form of the nerve root sleeve), they do not have the extra space and freedom of movement they had inside the dural sac. This makes them vulnerable to anything that would compress, stretch, pin or twist them - like a disc herniation. Not only that, but although the root sleeve is tough, it is far less tough than the layers of connective tissue that will protect peripheral nerves proper, once they are out of the spinal column and coursing down the leg.

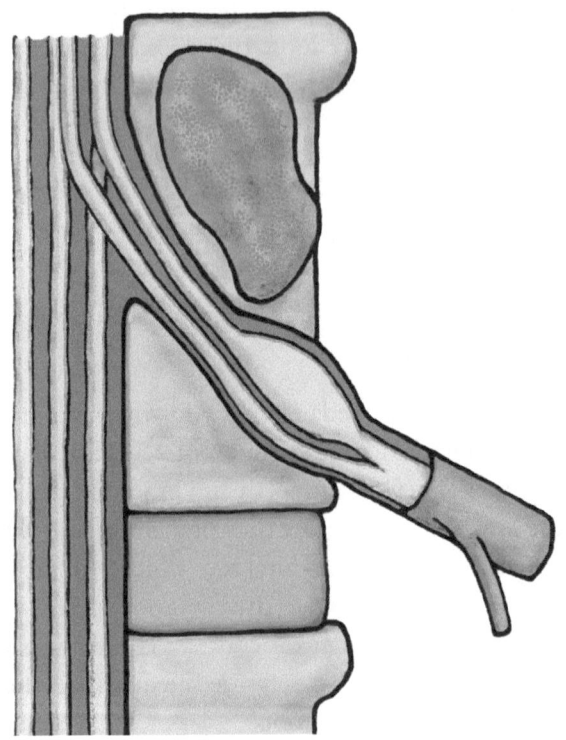

This picture shows the dorsal and ventral nerve roots
deviating from their downward course to leave the cauda
equina and the dural sac. They take with them part of the
dural sac which is now called the nerve root sleeve, shown in
dark grey. (Somewhat confusingly, when a sensory and a
motor nerve root are bound together like this, we still
colloquially refer to them as 'the nerve root', i.e. as a singular
when in fact they are distinct entities.) The dorsal root is
continuous with the dorsal root ganglion. Shortly after, the
roots blend to become the mixed spinal nerve. Before too
long, they branch into the ventral and dorsal rami.

This short interval in which the dorsal and ventral roots are outside
the dural sac but still inside the spinal column is called the radicular
canal. It can be thought of as a passageway, with the door at the end
being the opening of the intervertebral foramen. To pass through the
canal, the roots, which have so far been descending in a straight line

downwards, have had to swoop laterally as if exiting on a sharp slip road (or off ramp, for American readers!). Emphasising this swooping motion, they hug the ceiling of the passageway within the foramen.

The dorsal root ganglion (usually) sits in the intervertebral foramen

Inside the radicular canal, the dorsal (sensory) roots exit from the dorsal root ganglia. (The ventral roots, which have their cell bodies within the ventral horn of the spinal cord, course closely by). The root sleeve, which has protected the two roots since they left the dural sac, ends here. The dura mater of the sleeve blends into the capsule of the ganglion, soon to become the tougher connective tissue layers that protect the peripheral nervous system. The arachnoid membrane below it pinches off just before it reaches the dorsal root ganglion. This pinching cuts off the CSF, which means this point marks the last outpost of the central nervous system and everything beyond it is the peripheral nervous system.

The dorsal root ganglion is usually located in the foramen, sometimes slightly distal or proximal to it. In the lumbar spine, it is about the size of the fingernail on your little finger. Inside it are the cell bodies of just over 10,000 sensory neurons, each one with a diameter of less than 100 micrometres, which is about the width of a strand of hair. These cell bodies manufacture all the parts that the primary sensory neuron needs in order to function, and ship them out to the rest of the cell. They sit off from the rest of the primary sensory neuron at the end of a T-junction.

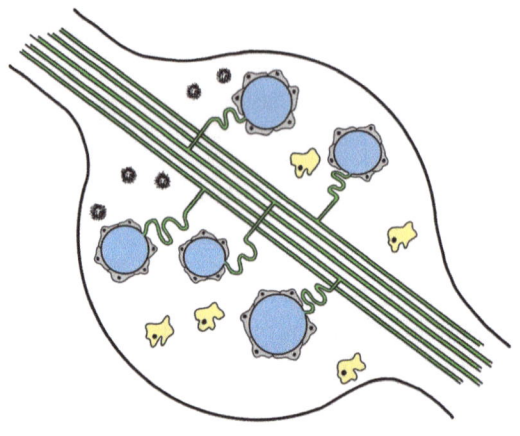

Schematic diagram of the dorsal root ganglion. Axons, green,
pass through. They send off T-junctions to cell bodies, blue.
Cell bodies are surrounded by satellite cells, grey. Resident
macrophages shown in yellow. T-cells cluster to the CNS-side
of the ganglion.

The cell bodies of sensory neurons are not the only cells inside the
dorsal root ganglia. For instance, satellite glial cells surround each cell
body. These glial cells have a protective and cushioning function akin
to that of the Schwann cells in the peripheral nerve trunk. They also
closely interact and communicate with neurons using, for instance,
immune mediators. Additionally, the dorsal root ganglion also
contains immune cells such as macrophages (15).

Although the ganglion is thicker than the nerve root, it still does not
take up more than one third of the foramen. The rest of the space
inside the foramen is taken up by blood vessels, ligaments, the sinu-
vertebral nerve (heading back into the canal) and cushioning fat.
Below, you can see two drawings of the teardrop-shaped intervertebral foramen.

Distal to the dorsal root ganglion, the nerve roots become the mixed spinal nerve

Distal to the dorsal root ganglia, the pair of nerve roots, up until now held closely but separately, undergo a major change: they are finally woven together into the mixed spinal nerve. This means that the spinal nerve contains both the motor and sensory neurons of its spinal level.

Compared to the length of the lumbar nerve roots, the lumbar spinal nerve is remarkably short: as soon as it has left the foramen, it branches off into two rami. The dorsal and ventral rami are mixed sensory and motor nerves that serve the structures of the spine and the lower limb, respectively. We have now left the nerve roots and the dorsal root ganglion behind.

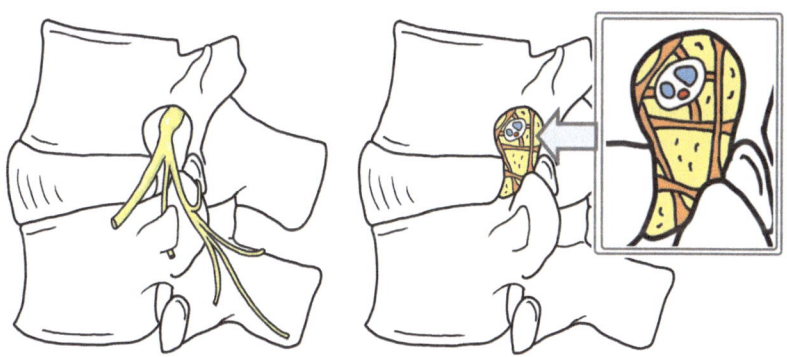

Two views of the intervertebral foramen. The picture on the left shows the dorsal root ganglion blending into the mixed spinal nerve which soon splits into the dorsal and ventral rami. The picture on the right shows a cross section of the nerve root (blue). Note the dorsal and slightly smaller ventral root, accompanied by the distal radicular artery (red) in relation to the latticework of intra-foraminal ligaments (orange). Note also the fatty tissue (yellow) which takes up much of the rest of the space. The fact that the foramen houses these other tissues enables nerve root irritation without direct contact, as we will see later.

The nerve roots are supplied by the radicular arteries.

Let's pause here to look at how the nerve roots and the ganglion get their blood supply, because the blood supply is one of the most under-rated factors in radicular pain!

Making its way down from the heart, the descending aorta sits on the front of the vertebral bodies. At each spinal level, a pair of arteries branch off from the aorta and course backwards around each side of the vertebra. They look as if they are a pair of arms hugging the vertebral body. As they make their way backwards, these arteries send branches off to various parts of the vertebra, and one of these branches enters the intervertebral foramen. This is the distal radicular artery. As it enters the foramen, the distal radicular artery penetrates the

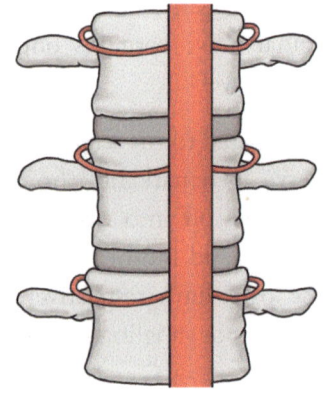

A view from the front of the vertebral bodies. At each spinal level, the descending aorta sends out a pair of arteries.

spinal nerve and follows it in, in the opposite direction we took on our tour. When it reaches the nerve roots, it splits in two and continues to course up both. On its way past the dorsal root ganglion, it forms a plexus around it.

As it continues, it gives off more collateral branches which become tiny capillaries in each bundle of neurons inside the root. It ends about two thirds of the way up the roots. It is through this distal radicular artery that the heart pumps blood up the nerve roots, towards the spinal cord.

The proximal radicular artery begins in the conus medullaris and travels down the roots. At first it travels alone, because capillaries from the conus can supply the most proximal parts of the roots. But after a few millimetres, the proximal radicular artery swerves and pierces the root. Like the distal radicular artery, it gives off collateral branches as it goes, which become tiny capillaries in each bundle of neurons. It is through the proximal radicular artery and its branches that the heart pumps blood down the nerves, away from the spinal cord.

The proximal radicular artery meets the distal radicular artery about one third of the way down the root, and the two blend into one another, their currents mingling.

Blood is drained from the root by radicular veins. Inside the nerve root, the veins run in spirals. Outside, they form large, thick plexuses throughout the radicular canal. The radicular veins drain de-oxygenated blood posteriorly, towards the spinous processes; the opposite direction from which it came. From there, it will continue to the lungs to be re-oxygenated.

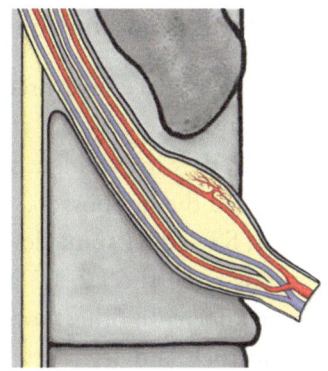

Blood supply to the nerve root complex. The distal radicular arteries (red) enter the nerve roots through the spinal nerve, on the right. The proximal radicular arteries arrive from the left, having come from the conus medullaris, out of shot. Blood flow from each converges at a watershed about two thirds of the way up the root. Veins shown in blue.

The nerve root complex.

We are about to leave the spine, but before we do, let's pause one more time, look back and introduce a new term to describe much of what we have just seen: the nerve root complex. This term is used by Butler in his 2000 book The Sensitive Nervous System (16), and he writes that he took it from Rauschning (17), who was writing a couple of decades prior. You can also read Sunderland talking about the 'radicular complex' in 1974 (18). We will use 'nerve root complex' to mean the dorsal and ventral nerve roots, the dorsal root ganglion, the nerve root sleeve and the radicular veins and arteries. It's a useful term when you want to refer to this collection of anatomically and physiologically different structures which bundle into a coherent whole, and all have their role to play in radiculopathy and radicular pain.

The plexuses become the femoral and sciatic nerves.

After the spinal nerve branches off into rami, the neurons in the ventral ramus travel off to the buttock and leg. Before they do, however, they intermingle neurons from other spinal levels in a structure called the lumbosacral plexus. Specifically, neurons from the L4, L5, S1, S2 and S3 spinal level combine to form the sciatic nerve, and neurons from the L2, L3 and L4 spinal level combine to form the femoral nerve. In other words, by the time our sensory neurons are in the buttock and leg, they are thoroughly jumbled up, each nerve contains neurons from many different spinal levels. This is why a radiculopathy, i.e. an injury at the nerve root, rarely causes complete strength loss: by the time the neurons that have been injured at the nerve root connect with muscle in the leg, they are working alongside uninjured neurons from other spinal nerve root levels. Typically, those uninjured neurons maintain some strength without their injured colleague.

(By the way, the name for each nerve root and spinal nerve is taken from one of the vertebrae where it exits. In the lumbar spine, a nerve is named for the vertebra above it. So, if we look at the intervertebral foramen between L4 and L5 we will see the L4 nerve roots and L4 spinal nerve. This rule is simple, but there is a source of confusion: because of a mismatch between the number of nerve roots and the number of levels, in the cervical spine we name them the other way round, after the vertebra below.)

Beyond the rami and the plexuses, nerves containing sensory and motor neurons from, mostly, multiple spinal levels, proceed into the buttock and leg. Motor neurons end in muscles. Sensory neurons originate from everything we can feel: muscles, bones, ligaments, cartilage, connective tissue, the skin, and the follicle of the smallest hair. If we look closely at the tips of these sensory neurons, we will see that they either form a receptor (e.g. a Meissner or Pacinian corpuscle) or have free nerve endings which are covered with channels.

Sensory receptors and ion channels start action potentials.

At the tips of sensory neurons, there are sensory receptors and ion channels, which work together to convert stimulations into electrical signals called action potentials.

Ion channels are proteins embedded in the membrane of sensory neurons that allow ions to pass through, into and out of the neuron. There are lots of different types of ion channels, and you don't need to know them individually to understand this book. But here are a few examples that you might have heard of. There's TRPV1 channels, which react to thermal stimuli, including sunburn. There's acid sensing ion channels, which are activated by a drop in tissue pH during ischemia. And there are voltage gated sodium channels, which react to a change in membrane potential to create and transmit action potentials. In addition to ion channels, there's receptors such as cytokine receptors, which are activated by tissue inflammation. But, as we say, you don't need to remember each one, and it's sufficient to think of the ion channels and receptors as a group - the cell machinery that allows a sensory neuron to 'feel' what's going on around it.

Ion channels and receptors are not born in the neuron tips (where they mostly reside) but, as we saw earlier, way up in the dorsal root ganglion. To get down to the neuron tips, they are actively carried along microtubules, a process called axonal transport. That means that the axon of a nerve is not an inert wire waiting for impulses; it is also a conveyor belt for the machinery that makes the cell work. This will become important later as we watch how radicular pain develops: as we will see, ion channels' and receptors' journey along an axon can be interrupted by an injury or inflammation, which makes them bunch up into an abnormally sensitive spot, contributing to neuropathic pain. Not only that, but after an injury to a primary sensory neuron, ion channels and receptors are also produced and shipped out along the axon in large numbers, which can make the neuron hyperexcitable.

And with that, we have finished our tour. We are down in the tissues, watching action potentials spark off in a primary sensory neuron's tips. They will zip up the neuron's axon, through the nerve root complex and up to the spinal cord and brain, in the opposite direction to that we've just traveled.

Now, you've seen everything you need to see to start building your understanding of sciatica.

Key points from the anatomy tour:

- The lumbosacral primary sensory neurons synapse with the spinal cord in the upper lumbar spine, and have their nerve terminals in the tissues of the pelvis, buttocks, legs and feet. The nerve root is just a short length of a bundle of these very long neurons.
- The lumbosacral nerve roots bud off from their respective spinal cord segment and travel down the lumbar spine as the cauda equina, where they are protected by the dural sac.
- At each spinal level, two pairs of sensory and motor nerve roots exit the dural sac, one pair on the left and one pair on the right. As they go, they take some of the dural sac with them as the nerve root sleeve.
- The nerve roots travel a short distance into the foramen. Around here, the dorsal nerve root has its ganglion, which

houses the cell bodies of the sensory neurons in the nerve root.

- Distal to the dorsal root ganglion, the sensory and motor nerve roots blend together into the mixed spinal nerve. This is the point at which the nerve root ends.
- The spinal nerve is very short, and the neurons inside soon branch off to innervate the lower back and leg.
- The nerve roots are supplied by the radicular arteries.
- The extra-thecal part of the nerve root, the dorsal root ganglion, the mixed spinal nerve, the nerve root sleeve and the radicular arteries can usefully be described as 'the nerve root complex'.
- Sensory neurons innervate all tissues that we can feel. In their nerve endings, they have ion channels and specialised receptors. These channels and receptors convert stimuli like touch, heat, cold, and chemical changes into action potentials, which are sent up to the spinal cord and may eventually reach the brain.
- These ion channels and receptors are manufactured in the dorsal root ganglion and transported to the sensory nerve endings by axonal transport. As we will see, problems with the manufacturing and transport of these channels and receptors is a significant mechanism of radicular pain.

PART II: THE NERVE

3

NERVE ROOTS CAN BE INJURED BY 1) MECHANICAL PRESSURE AND 2) CHEMICAL IRRITATION

'After all, it might be said that sciatica is hardly completely explained by the knowledge of disc protrusions...'

— *LENNART SÖDERBURG, 1956 (19)*

How pressure can injure nerve roots.

Mechanical pressure can injure nerve roots. But pressure means lots of different things - not just compression! To help paint the picture, here's an unofficial but hopefully intuitive typology of the three kinds of pressure that can injure nerve roots.

The most dramatic kind, and what most people think of when they picture a disc herniation injuring a nerve root, is squashing. Here are the researchers Smyth and Wright describing such a case: 'At operation there was a large disc protrusion pressing on the [...] nerve root, pushing it laterally. The root was wedged between the disc and the lateral wall of the spinal canal. Compression was severe' (20).

But squashing is probably rare compared to the second type of mechanical pressure, which is bowstringing (21).

Bowstringing happens because the root is relatively fixed in place against the disc side of the spinal canal by a set of ligaments. And, as we saw earlier, the root is further restricted by the tight swaddling of the root sleeve. Describing the fixity of the root, the neurosurgeon Murray Falconer wrote in 1943 that 'attempts to displace the [...] nerves are resisted, and if displaced the structure springs back into its original position when released' (22).

This fixity means that, in contrast to the roots inside the dural sac which can be compared to loosely-hanging ropes, once the roots are in the root sleeve they are more like guy ropes on a tent. And if you push your hand against the guy rope on a tent, it doesn't move much. Instead, it bowstrings. As it bowstrings, it distorts, stretching, thinning, and squashing in on itself. (Note that this is not simply 'compression'; the mechanical forces here are quite varied).

Nerve roots are not loose, but fixed at both ends like the guy rope on a tent

A disc herniation can have a similar effect on a nerve root. Observing this, Falconer described a 'compressing angulating as [the nerve root] was pulled taut and bent over the prolapse'. Here's Smyth and Wright describing the same thing: 'At operation there was a large dome-shaped herniation pressing on the [...] nerve root. The root was not compressed against the lamina.' This is bowstringing!

Finally, a nerve root can also be injured by mechanical pressure that is not as frank and direct as squashing or bowstringing. It might not even involve nerve root contact. This third type of pressure is 'crowding out'. As well as disc herniations, crowding-out can be caused by changes like osteophytes, ligament thickening, disc flattening, disc bulging or even oedema from inflammation or venous congestion (more on that later).

How can crowding-out injure a nerve if the nerve itself is not deformed and perhaps not even contacted? Well, the nerve root shares its space in the radicular canal with blood vessels, ligaments, fatty tissue and other nerves. So it can easily get tight in there. Think of standing in a crowded lift (or elevator!) and one more person gets on. As we will see soon, this little bit of added pressure on a root can be enough to affect its blood supply, causing a loss of nerve function, Schwann cell dysfunction and even inflammation.

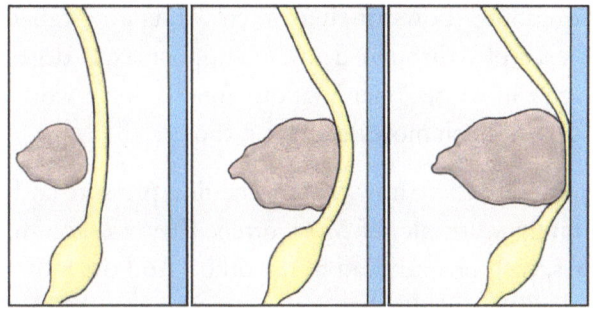

Schematic diagrams showing the different types of mechanical pressure that a disc, or any other intra-spinal mass, can exert on a nerve: crowding out, bowstringing and squashing.

How chemical irritation can injure nerve roots.

Sometimes this chemical irritation is introduced by surrounding structures. Most commonly, the culprit is disc material. Disc material chemically irritates a nerve root for two reasons. First, it can contain pro-inflammatory chemicals that contact neural tissue when the disc ruptures. Second, the innermost material of a disc is foreign to the immune system, so it can cause an autoimmune-inflammatory reaction that catches neural tissue in the cross-fire. Smyth and Wright might have been seeing disc-material-induced chemical irritation when they described this painful nerve root: 'At operation there was a small nodular herniation touching the fifth lumbar-nerve root. The root was quite free, there were no adhesions, nor did there appear to be any pressure' (20).

There are other surrounding structures that can introduce chemical irritation to a nerve root. For example, facet joint inflammatory chemicals can irritate the root (23). Malignant tumours release inflammatory chemicals, too (24).

Sometimes, chemical irritation is not introduced by a surrounding structure but arises as a response to mechanical pressure. Pressure, whether squashing, bowstringing or crowding-out, can damage a nerve (for example, through demyelination or axon degeneration). And damage can cause inflammation. Again, don't worry, we will look at all this in much more detail later, too!

You will have noticed by now that mechanical pressure and chemical irritation rarely work alone. Most often, they are accomplices. In various ways, each one can lead to the other. And the most common cause of radicular pain, disc herniations, can be a cause of both simultaneously.

Nevertheless, it helps to split them up into these two categories. It's a good way to make sense of things. For the rest of the first half of this book, we are going to use mechanical pressure and chemical irritation to understand how a nerve root can lose function, and how it can induce pain.

Key points for nerve root injuries

- Mechanical pressure can injure nerve roots.
- There are different types of mechanical pressure. Squashing is when a nerve root is compressed against something. Bowstringing is when a nerve root is displaced along its course. Crowding-out is when a nerve root is not necessarily

contacted physically by tissue, but is indirectly mechanically disturbed.

- Chemical irritation can injure nerve roots
- There are different kinds of chemical irritation. Disc material is a chemical irritant to nerve roots. Inflammation from nearby joints can irritate nerve roots. Pressure itself can cause an inflammatory reaction inside nerve roots. Ischaemia can also irritate a nerve.
- In reality, mechanical pressure and chemical irritation overlap, but looking at each one separately is a useful way to understand sciatica.

4

HOW MECHANICAL PRESSURE CAUSES RADICULOPATHY

'The basic effect of the compression is a degeneration of a greater or lesser number of the nerve fibres [...] the degenerating nerve fibres are usually diffusely strewn all over the cross section in smaller or greater numbers in proportion to the severity and duration of the compression [...] As soon as a nerve fibre begins to degenerate there occur reactive regenerative processes...'

— LINDBLOM & REXED, 1948 (25)

When tissues are squashed and distorted, blood cannot flow in. For example, if you pinch or press the skin of your hand, it will blanch for a split second, then return to its normal colour when blood flow is restored. Nerves are the same: add enough pressure, and you cut off the blood.

Let's watch what happens when a neuron is deprived of blood.

Picture the short stretch of a sensory axon as it passes through the nerve root. An action potential arrives from the big toe. This stretch of the nerve root passes on the action potential by opening its voltage gated sodium channels (a class of those ion channels we saw earlier) and allowing sodium ions to rush in, raising the charge across its membrane. Crucially, once the action potential has passed, the neuron needs to restore the charge across its membrane to a resting state so that it is ready to conduct the next impulse - i.e., it needs to reset. To do this, the neuron has to transport those sodium ions back out of the axon, and transport potassium in. This repolarisation requires oxygen. And oxygen is borne in the blood.

That means that if this nerve root segment does not have blood, it has no oxygen to restore the charge across its membrane. And if it cannot restore the charge, the next action potential that arrives will fizzle out and go nowhere. When a stretch of a neuron cannot conduct action potentials, we call this a conduction block.

So, adding mechanical pressure to nerves reduces blood flow which leads to a conduction block. This is an everyday experience for most people. Have you ever sat with your legs folded underneath you and your foot has turned numb and gone limp? Sitting in such a way reduces the blood supply to your common peroneal nerve until it does not have enough oxygen to restore its membrane potential and it can no longer conduct action potentials.

Just like in the peroneal nerve, pressure on the nerve root complex (whether it's squashing, bowstringing or crowding-out) can also reduce blood flow. If blood flow is reduced enough to cause a conduction block, then we will see the numbness and weakness we call radiculopathy.

Let's look more closely at how that happens.

* * *

Venous congestion can lead to a conduction block

'Even moderately increased pressure may significantly affect a peripheral nerve'

— GÖRAN LUNDBORG (26)

A surprisingly small amount of added pressure is capable of causing a conduction block in a nerve - given enough time.

For example, in one study, just 30mmHg of added pressure on the median nerve as it runs through the carpal tunnel was enough to stop sensory neurons from responding after an hour, and cause pins and needles in the hand (27). This is well below mean arterial pressure, which is around 70-100mmHg. Although it's challenging to replicate such studies in humans' nerve roots, it's likely that nerve roots are, if anything, less equipped to withstand pressure than median nerves. Supporting this, a study in the nerve roots of pigs supports the idea that a relatively small amount of added pressure can impede blood flow through the nerve root complex enough to cause a conduction block (28). And one study with people undergoing discectomy found that the average amount of pressure exerted by a disc herniation on a nerve root was, at the time of operation, a modest 53mmHg (29).

Why can a relatively small amount of added pressure be so detrimental?

Well, for blood to flow into and then out of a nerve, there must be a particular pressure gradient (30). Blood is pumped in through the arterioles, which have the highest pressure. Then it continues into the capillaries in the nerve fascicles which have a lower pressure than the arterioles. Then, after giving up its oxygen, the blood drains into the venules, which have a lower pressure still. And the lowest pressure of all must be in the space surrounding the nerve, so that deoxygenated blood can flow out of the nerve again. If this pressure gradient is maintained, then blood can flow.

But if there is additional pressure on the nerve root complex, then the normal pressure gradient could be reversed, causing a backlog. And that additional pressure *only needs to be greater than the lowest pressure in the gradient to cause that backlog.* To put it another way, if the pressure in the space surrounding the nerve is greater than the pressure in the venules, then deoxygenated blood cannot easily flow through those venules and out of the nerve root complex. And if deoxygenated blood cannot drain out, then oxygenated blood cannot be pumped in through the arterioles. The proper name for this backlog is *venous congestion.*

So, that's how a relatively small amount of added pressure can cause a conduction block in a nerve root: by stopping blood from flowing at the lowest-pressure end of the gradient.

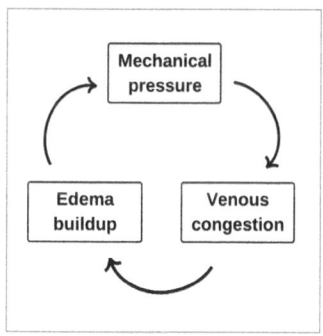

There's more! This venous congestion can start off a vicious cycle. The backlog causes fluid from congested capillaries to leak out through the blood-nerve barrier into the surrounding nerve root (31). This is called edema. And edema only adds to the already raised pressure in the area. This further occludes circulation. The researcher Göran Lundborg pointed out that in effect, this vicious cycle is 'a compartment syndrome in miniature' in the nerve (26).

Unfortunately, edema builds up particularly easily in nerve roots. This is because roots do not have the tight blood-nerve barrier of peripheral nerves, which makes it easier for fluid to leak out from the capillaries (32). And roots also have the tight swaddling of the nerve root sleeve, which stops edema from dispersing. In animal experiments, researchers observe edematous nerve root thickening that is visible to the naked eye. And surgeons often describe, as Grieve put it, 'grossly edematous, hyperaemic nerve roots' (33).

In fact, the buildup of edema and subsequent vicious cycle is not confined to the area of the nerve under pressure (34). Falconer observed that 'This edema can extend some distance up and down the nerve from the point of compression [...] we have frequently seen at operation marked swelling of components of the affected nerve root within the dural sac' (22).

So, venous congestion and edema buildup are sufficient to cause a conduction block in the nerve by depriving it of blood and therefore oxygen. This kind of conduction block is sometimes called a *reversible* conduction block because, as long as there is no underlying structural damage to the nerve, it only lasts as long as the venous congestion is maintained. For example, some people recover their strength and sensation very quickly after lumbar decompression surgery (and after carpal tunnel decompression surgery, too). Imaging suggests this is probably because once pressure is relieved, edema can disperse and the blood-nerve barrier can recover (34,35). A reversible conduction block also accounts for the cases where someone with spinal stenosis can voluntarily bring on their numbness and weakness by extending their spine and ease it by flexing. Elsewhere in the body, another good example of a position-dependent reversible conduction block is carpal tunnel syndrome: think of Phalen's test, which increases pressure in the carpal tunnel to elicit symptoms.

Here is another example of a reversible conduction block caused by a lack of oxygen to a nerve root. Earlier, we mentioned the research of Smyth and Wright. As part of their investigations, they tied nylon thread around the nerve roots of some patients and, if you can believe it, left the string in place, with the ends hanging out from the body, after the operation (20). After a few days, they would pull gently on the string to see what happened. Often, prolonged gentle pressure on a nerve caused a steadily increasing reversible conduction block that eased when the researchers stopped pulling. As Smyth and Wright wrote of one case, 'After a continuous but gentle pull for a quarter of a minute, the patient volunteered the information that all of the toes and part of the dorsum of the foot, proximally, felt numb'. Presumably, since Smyth and Wright pulled on the threads gently, this is a reversible conduction block caused by venous congestion and the fifteen seconds is the time it took, in this case, for the backlog to establish.

To finish this section, a quick word on the dorsal root ganglion. The ganglion needs even more oxygen than the rest of the neuron, so it can really suffer when the circulation to the nerve root complex wanes. On top of that, the ganglion has no blood-nerve barrier, which means that when venous congestion sets in, edema leaks out of the blood vessels and into it the ganglion more easily. And once there, edema cannot easily escape from of the ganglion into the surrounding tissues because of the ganglion's tight capsule. In one experiment, compression of a dog's ganglion for anything longer than 20 minutes caused a lasting effect on the function of the neurons inside (36).

Ischaemia can lead to an often irreversible conduction block

We have seen how raised pressure makes it hard for blood to flow out of a nerve, which causes venous congestion. But, as the pressure around a nerve gets higher still, it becomes increasingly hard for blood to flow into the nerve in the first place. This is called ischaemia.

Ischaemia causes more serious and lasting damage to nerves. Under conditions of venous congestion, axons do not have enough oxygen to function but there is still some oxygenated blood coming in and flowing through. But under conditions of ischaemia, no oxygenated blood is coming in at all. This means that not only can a nerve no longer conduct action potentials but it also cannot support the health of its myelin sheaths and its axons. Structural damage ensues. Confirming this, one systematic review of animal models found that experimentally-compressed nerve roots can recover rapidly after decompression unless that compression was enough to exceed mean arterial blood pressure, causing ischaemia, in which they recovered very slowly or not at all (37). To put it another way, under conditions of ischaemia, nerve roots were much more likely to suffer an irreversible conduction block.

Irreversible conduction blocks are common for people with radiculopathy. In one study, for example, 20% of people still had a conduction block three months after decompression surgery (38). The extent of recovery depends on the extent of the damage. If it is myelin loss only, then nerves can re-myelinate in weeks to months. If it is axon loss too, then it will take longer.

Of course, whether the nerve can recover also depends on whether the blood supply is restored after the nerve is released from the added pressure. And that is not guaranteed. In his experiments, Rauschning found that 'in severely degenerated lumbar spine segments, no contrast medium could be forced into the arterial branches inside the root canals... this phenomenon may be explained by [...] hypotrophy of the blood vessels resulting from long-standing degenerative encroachment' (39). This might explain why a minority of people, perhaps 10% (38,40,41), still have a radiculopathy a full year after its onset, even after decompression surgery (42).

Let's look a bit more closely at axon loss. Under mechanical pressure, axons do not only degenerate at the site of pressure, but along their length, too. However, they degenerate on one side of the pressure only: the side that's cut off from the cell body (43). As we have seen, a cell body supplies the entire neuron with the parts it needs to function, and receives used parts from the neuron for waste disposal. So, the part of an axon that is still in contact with the cell body has what it needs to survive. The part that's cut off from the cell body does not.

This 'dark side of the neuron' is left high and dry, and gradually degenerates. This is called Wallerian degeneration. (An analogy for Wallerian degeneration and subsequent regeneration is with a salamander that has lost its tail: the part with the head can survive and regrow a new tail, but the lost tail itself will die.)

When it comes to the nerve root, pressure can cause Wallerian degeneration in two directions. Sensory neurons degenerate in the direction of the spinal cord, away from cell bodies in the dorsal root ganglion (44). And motor nerves degenerate towards the periphery, away from their cell bodies which are in the spinal cord. This has been observed in nerve roots by Kobayashi and colleagues who described patches of degenerated nerves, swollen with edema and infiltrated by macrophages, in just such a pattern (44). Some people think the direction of Wallerian degeneration in nerve roots might be one reason why motor function recovers more quickly than sensory function after a nerve root injury (45): motor roots need to recover in the sturdy and straightforwardly-organised peripheral nerve, whereas sensory roots need to recover at the complex junction of root and spinal cord (44). This dorsal root entry zone seems to be particularly hostile to regenerating sensory axons (46).

* * *

Schematic representations of the normal and ailing circulation

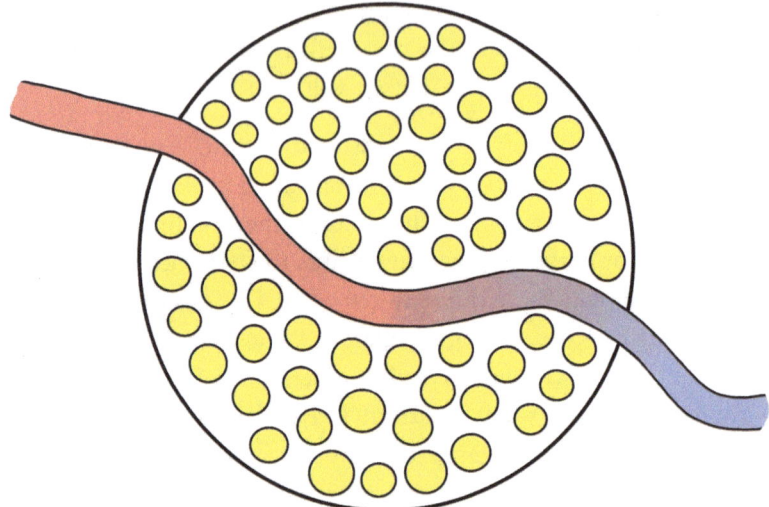

Normal circulation, cross section through a nerve root. Individual neurons in yellow. For blood to flow into and then out of a nerve, there must be a particular pressure gradient. Oxygenated blood (red) is pumped in through the arterioles, which have the highest pressure. Then it continues into the capillaries in the nerve fascicles which have a lower pressure than the arterioles. Then, after giving up its oxygen, deoxygenated blood (blue) drains into the venules, which have a lower pressure still. And the lowest pressure of all must be in the space surrounding the nerve. If this pressure gradient is maintained, then blood can flow.

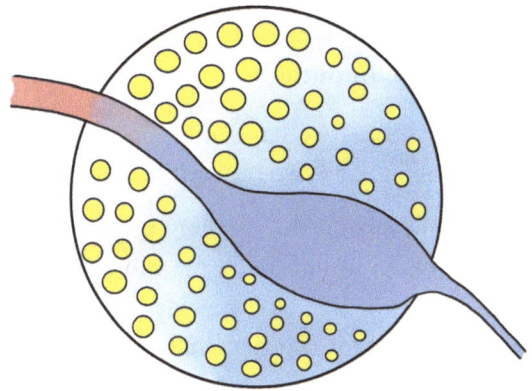

Venous congestion. When the pressure in the space surrounding the nerve is greater than the pressure in the venules, then deoxygenated blood (blue) cannot easily drain out of the nerve root complex through those venules. And if deoxygenated blood cannot drain out, then oxygenated blood (red) cannot be pumped in through the arterioles. This creates a backlog called venous congestion. Note that edema has leaked out and compressed some of the neurons on the right of the image.

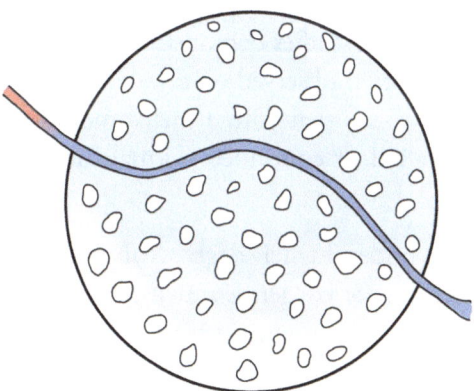

Ischaemia. As the pressure around a nerve gets higher still, it becomes increasingly hard for blood to flow into the nerve in the first place. The nerve is deprived of oxygen and neurons degenerate. This is called ischaemia.

As well as the degree of pressure, time under pressure and speed of onset are also important

So far, for simplicity's sake, we have only focused on the *degree* of pressure. Now, we need to add two more things to the picture.

The first is the amount of *time* the nerve root is under pressure. After all, if you sit on your peroneal nerve for ten minutes, your foot might go limp and numb but it will recover quickly (a reversible conduction block). But people with radiculopathy may have added pressure on their nerve roots for weeks or months, so that, even after decompression, recovery also takes months (an irreversible, or at least very slowly-reversing, conduction block).

Time matters because many of the downstream consequences of pressure, such as the build-up of venous congestion, demyelination and an inflammatory response, take hours, days or weeks to play out. For example, Yoshizawa and colleagues applied a loose band around dogs' nerve roots, 'slightly larger than the diameter of the root', and didn't observe any nerve degeneration or a conduction block for three months (31). Many other studies demonstrate that pressure-induced changes in nerves are often observed on a scale of weeks and months (47). In fact, it's fair to say that with a small amount of added pressure on a root, a period of continued normal function is expected (48).

Time also matters under higher degrees of pressure because, of course, a prolonged loss of oxygen is much more harmful to neural tissue than transient loss of oxygen (49–51). The time factor is why cauda equina syndrome is an emergency, and why early surgery is better than later surgery for recovery of nerve function for people with radicular pain (52). In fact, the amount of time a nerve is under pressure is probably just as important as the degree of pressure (37).

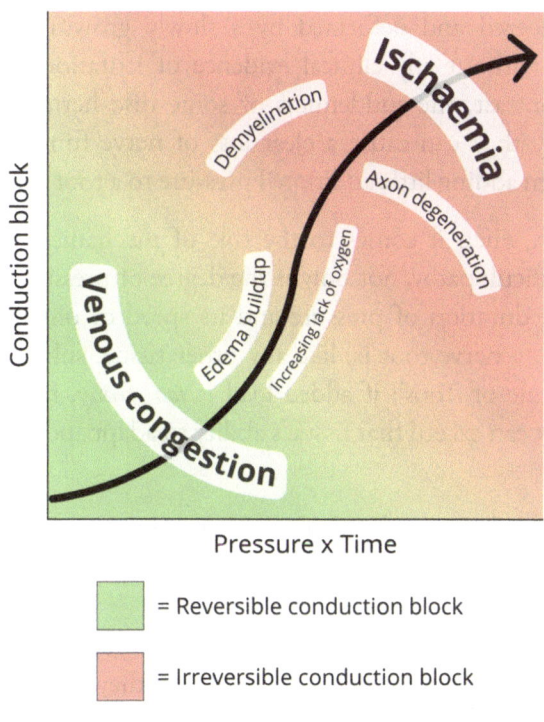

Graphical representation of the effect of pressure and time on
nerve physiology and conduction

Along with time, the other thing we need to add to the picture is the
speed of onset of the pressure. In pigs, slowly increasing pressure over a
number of weeks causes less venous congestion than applying pressure all at once (28). In one experiment, even a difference of seconds
affected the amount of edema build-up (53). It seems that, given
time, the nerve root complex is can adapt (54). This is one reason why
many older people with progressive foraminal narrowing can have, as
Grieve described it, 'flattened, almost fenestrated' nerve roots but

never have any symptoms (33). Grieve also quotes Frykholm, who wrote in 1971 that 'it is quite amazing to what extent a nerve root can become squeezed and deformed by a slowly growing osteophytic protrusion, without any clinical evidence of irritation or dysfunction'. By contrast, the suddenness of some disc herniations partly explains why they can cause a clear loss of nerve function, despite many of them adding little additional pressure to a root complex.

In summary, when it comes to the role of mechanical pressure in causing a radiculopathy, not only is the degree of pressure important but also the duration of pressure and its speed of onset. To put it another way, a nerve root is, like any other tissue, subject to David Poulter's 'Rule of Toos': if added load is *too strong*, *too long* or *too quick*, then it can exceed that tissue's ability to adapt and cause injury.

Key points on how mechanical pressure can cause a radiculopathy:

- Mechanical pressure can cause a conduction block in a nerve root by depriving the axons of the oxygen they need in order to function.
- The effects of pressure can be split into two:

  1. A relatively small amount of added pressure can deprive the root of oxygen by creating a backlog in the root's local circulation. This is called venous congestion. Provided it is not too prolonged, this type of conduction block will often reverse rapidly when pressure is released.
  2. A relatively large amount of added pressure can stop blood from flowing into the nerve root in the first place. This is called ischaemia.

- Although venous congestion and ischaemia are on a spectrum and overlap in their effects, on the whole ischaemia causes greater structural nerve injury, principally in the form of demyelination or axon degeneration. This type of conduction block is unlikely to immediately reverse even once the pressure is released.

- Although the degree of added pressure is important, perhaps equally important are 1) the amount of time the pressure is exerted, 2) and the speed of its onset.

5

SIDE NOTE: FIBROBLASTS AND SCAR TISSUE

'Sometimes the ganglion and nerve adhered to the protruding disc by dense connective tissue, so that their separation from the disc had to be made with a knife...'

— LINDBLOM & REXED, 1948 (25)

Let's take a moment to look at something a bit different. We have just seen how venous congestion and the resulting edema buildup can cause a conduction block. But another consequence of venous congestion and edema build up is that fibroblast cells lay down scar tissue, making the nerve and its sleeve fibrotic (55). In one of the first studies to identify fibrosis, Hoyland and colleagues wrote that 'in almost every case, neural fibrosis was associated with a reduction in the quantity of neural tissue' (56). These changes are likely to be enduring.

As well as within nerves, fibrosis can also develop between nerves and nearby tissues, causing them to adhere. Here are Smyth and Wright, describing such a case: 'There was a small nodular disc protrusion pressing on the fifth lumbar-nerve root. Its apex lay at the centre of the anterior part of the nerve and the latter was bound to the protruding disc by a number of adhesions' (20). This adherence of a nerve root to nearby tissues is sometimes called tethering. It holds nerve roots in place, stopping them from moving freely. With movement, this increases pressure on roots, which might drive further pain and loss of function. (Although it's also worth noting that the fibroblasts that cause fibrosis and adhesions are themselves capable of directly sensitising a nerve (57). Again, mechanical pressure and chemical irritation are linked, not distinct.)

Illustrating the role of adhesions, Kobayashi and colleagues found adhesions between herniated discs and nerve roots in patients undergoing microdiscectomy for sciatica (43). After freeing up the adhesions and removing the hernia, their intra-operative nerve root movements immediately improved, as did the post-operative straight leg raise range of motion. It is unlikely that adhesions are the only reason for limited straight leg raises, but they may play some part.

Fibroblasts also lay down scar tissue after surgery. For example, here is the surgeon Stephen Kuslich, who conducted a study in which he stimulated the tissues of awake patients intra-operatively: '[In] patients who had undergone prior laminectomies [...] there was always some degree of perineural fibrosis. The scar tissue itself was never tender. The nerve root, however, was frequently very sensitive' (58).

It is not clear how important fibrosis is for radiculopathy and radicular pain. One systematic review found that operations to break down fibrosis are effective for the pain (59). But the association of fibrosis with pain in cohort studies is inconsistent (60). It's an understudied area.

Finally, on rare occasions, nerve root adhesions might cause a marked, mechanical restriction of movement with relatively little pain.

McKenzie first identified this phenomenon and named it adherent nerve root syndrome. According to McKenzie, people with adherent nerve root syndrome are almost always young and have leg pain only at end range spinal flexion, straight leg raise and slump positions, all of which are very limited.

6

HOW CHEMICAL IRRITATION CAUSES RADICULOPATHY

Let's look at how chemical irritation causes a nerve conduction block.

First of all, where does this inflammation come from? Well, as we mentioned earlier, disc material is a chemical irritant to a nerve root. We will look more closely at exactly why when we look more closely at the disc. But it is partly because disc material contains its own pro-inflammatory chemicals, and partly because when it herniates, it causes a local auto-immune inflammatory reaction.

Olmarker and colleagues were the first to show that disc material is a chemical irritant (61). They took a group of pigs and applied disc material, specifically the nucleus, to their nerve roots. They made sure that the disc material merely touched the root, without substantial pressure. In a control group, they applied fat to the nerve instead. The researchers found that the disc material caused the pigs' neurons to conduct impulses more slowly: a conduction block.

Not only disc material, but pressure itself - whether from a disc, age related changes, a benign tumour or anything else - can also cause inflammation in the nerves. After all, inflammation is common whenever the homeostasis of a tissue is threatened. This was shown in nerve roots by Kobayashi and colleagues, who clipped the nerve roots of dogs and observed that the nerves became infiltrated with macrophages and mast cells (44).

Finally, there are other causes of local inflammation that can 'catch' the nerve root and cause problems. For example, it's been shown that facet joint inflammation can reach nerve roots (23).

Macrophages and mast cells! Maybe these words make your eyes glaze over. Maybe you have tried reading about inflammation before and know what's coming next: T cells, leukocytes, cytokines, TNFα, IL1 β, BDNF, TLR4, YMCA... So let's try to break it all down.

Inflammation is the immune system's response to anything that it thinks might be dangerous. 'Dangerous' means injury and infection, of course, but also many of the other stresses and strains of being alive, being unhealthy and getting old. In nerve roots, the immune system is most likely to kick off an inflammatory reaction in response to the perceived dangers of disc material, added pressure from any of the surrounding structures, or hypoxia.

The immune system is run by immune cells. Here's a few of the immune cells that are in inflamed nerve root complexes: neutrophils, macrophages, mast cells, B-cells and T cells. Although all these cells do different things, we can get by just seeing them as a collective: the immune cells. A few of them already live in the nerve root (they're called resident immune cells), steadily maintaining day to day homeostasis. They are the first to notice any trouble. Most live in the blood supply and hone in on the nerve root when they find out it is injured (they're called circulating immune cells) (62). And some are still in immature form in the lymphatic tissues or bone marrow, but can activate and enter the bloodstream if additional help is needed.

To do their job in the nerve root, the immune cells that are borne in the blood need to get access to the neurons. But, neurons have a tight blood-nerve barrier. Oxygen and nutrients can get through, but not big immune cells. To solve this problem, after nerve injury the lining of blood vessels becomes more permeable and the blood-nerve barrier breaks down (63). Now immune cells can infiltrate the nerve.

(As an aside: here is one example of how chemical irritation overlaps with mechanical pressure. When the blood-nerve barrier breaks down, fluid can enter the nerve more easily. So, by different mechanisms, both venous congestion and inflammation cause fluid to enter the nerve from blood vessels, and edema to build up!)

Once circulating immune cells join resident immune cells in the nerve, they get to work. They do two things. One thing they do is release inflammatory mediators; but we are going to leave the mediators for now and look more closely at them later. The other thing they do is break down injured, infected and stressed tissue. They do this by enveloping and ingesting the tissues or, if the tissues are too big, releasing damaging chemicals to do the job for them. In the nerve root, this means immune cells break down myelin sheaths and axons (64,65).

If we think back to the study we mentioned earlier, in which Olmarker and colleagues applied disc material to pigs' nerve roots and found that the roots conducted impulses more slowly, we can now fill in some of the blanks. Because disc material is a chemical irritant to nerve roots, immune cells rushed to the area, strolled through the broken-down blood-nerve barrier and got to work taking apart axons. Indeed, on closer inspection, the researchers also found that inside the pigs' neurons, the axons and Schwann cells were degenerated (61). In further work, Olmarker and colleagues confirmed that disc material makes the blood-nerve barrier of a nerve root become more permeable so that immune cells travel through it. Confirming that inflammation was the culprit, they also found that anti-inflammatory therapies protect pigs' nerve roots from the effects of disc material (66–68).

Disc material also causes chemical damage, without any additional pressure, to the dorsal root ganglion. In one experiment, disc material caused a crescent of inflammation to appear quickly around the capsule of the ganglion (69). After a few hours, the inflammation started to infiltrate the capsule. In another experiment, disc material induced cell death of the cell bodies in the ganglion after just 24 hours (70).

All this might seem a bit strange. Why would the body respond to danger by tearing itself up? It is because inflammation is an essential protective response. Degeneration precedes regeneration! An analogy might be a controlled forest fire. While it may seem destructive, the fire is essential for the health of the forest because it clears away dead and diseased trees to make room for new growth. Inflammation works in a similar manner, breaking down damaged tissue so that it can be replaced with healthy tissue.

Accordingly, if injured neurons are experimentally deprived of pro-inflammatory immune cells they simply do not regenerate (65). And there is evidence that a robust acute inflammatory response in fact protects against chronic pain (71).

In their review of inflammation, Antonelli and Kushner quote the physiologist Claude Bernard: 'All of the vital mechanisms, however varied they may be, have always but one goal, to maintain the uniformity of life in the internal environment' (72). Inflammation is no different.

Key points on how chemical irritation can cause a radiculopathy:

- Disc material seems to be inherently chemically irritating to a nerve root and this can cause it to become inflamed. Additionally, pressure itself can trigger an inflammatory reaction.

- As part of this inflammatory response, the blood-nerve barrier becomes more permeable and immune cells enter to break down axons and myelin in the nerve root. This can cause a conduction block.
- Although inflammation can be destructive, it is best seen as the immune system's attempt to heal a damaged area.

7

THE ROLE OF THE NERVE DISTAL TO THE NERVE ROOT COMPLEX IN RADICULOPATHY

As we have seen, injured axons undergo Wallerian degeneration; that is, they degenerate away from their cell bodies. For sensory neurons injured at the nerve root, this means their axons degenerate proximally towards the spinal cord.

It is somewhat surprising that when researchers take skin biopsies from the legs of people with chronic radicular pain, they find fewer nerve fibres there, too. One study found that in a group of people with radiculopathy, half of them had loss of intra-epidermal nerve fibres (73). Another study compared people with radiculopathy to people with a particularly severe form of neuropathic pain called complex regional pain syndrome and found that both groups had a similar amount of fibre loss in their skin (74). If the results from these two studies generalise, then it means that when you are testing someone's pinprick sensation and you find that part of their skin is numb, they might have a conduction block but it could also be a loss of their nerve fibres in the skin. Unfortunately, both studies were small, so it is hard to make strong conclusions.

It is not clear what causes peripheral cutaneous axon loss for people with radiculopathy. One possibility is that it has do with the cell bodies in the dorsal root ganglion. Recall that the cell bodies support the whole of the sensory neuron by sending parts up and down a long supply line. As the neuroscientist Marshall Devor wrote, 'the [cell body] should be seen as an incredibly busy little factory struggling full time to provide for the huge metabolic load of its enormous axon' (75). If the cell body is injured or dead, then it cannot send enough (or any) of these parts so the sensory neuron begins to degenerate too. That would explain distal axon loss.

However, just to make things more confusing, people with radiculopathy who have some nerve fibre loss in the skin of their affected leg also have some nerve fibre loss in their *non-affected* leg (73,74)! Such contralateral small fibre degeneration has also been observed in patients with complex regional pain syndrome (76). This, together with the frequent occurrence of small fibre loss in central disorders such post-stroke pain, suggests that the observed small fibre degeneration may not be fully explained by cell body death, but could be caused by more systemic factors. One culprit that has been proposed is systemic or remote inflammation (77), which we'll look more closely at later).

Researchers are still trying to find an answer to explain this observation. Either way, it is evident that some people with radiculopathy have nerve fibre loss in their skin, and sometimes also in the non-affected leg. This implies a neuron-wide and even system-wide problem that is not restricted to the section of the neuron that makes up the nerve root complex. That said, it's not clear yet whether all this affects the clinical decision making and choice of management strategies.

Key points on the role of the nerve distal to the root in radiculopathy

- People with radiculopathy can have intra-epidermal nerve fibre loss in the extremities.
- It's not clear why this is. It might be because the cell body is too injured to maintain its axon, or it might be because the whole cell has died. But the fact that these changes are sometimes also seen in the non-painful leg suggests there is a central or systemic cause.

8

SIDE NOTE: LOSS OF FUNCTION IN RADICULOPATHY IS USUALLY MILD

Before we finish looking at loss of nerve function, it's worth pointing out a key difference between radiculopathy and peripheral nerve lesions.

Peripheral nerve lesions can cause profound loss of nerve function. For example, radial nerve palsy can cause wrist drop, meralgia paraesthetica can cause profound numbness of the skin of the lateral thigh, severe carpal tunnel syndrome can cause true thenar muscle wasting, and even if you just sit with your legs crossed you can temporarily paralyse your own foot.

By contrast, radiculopathy usually causes a vague loss of sensation that patients might not even notice until you test them. If there is true numbness of the skin, it is usually a small patch. And radiculopathy seldom causes real paresis or muscle wasting.

Why? Radiculopathies usually cause a more mild loss of function because the neurons of the affected nerve root mix up with others in the lumbosacral plexus, so that by the time their fibres get to the skin and muscles as peripheral nerves, they are sharing their work with other nerve roots. So if there is a lesion at the nerve root, other nerve roots can pick up the slack.

By contrast, if there is a lesion at the peripheral nerve, that nerve is often the only thing that does its job. So it is analogous to cutting a cable on a TV: the TV goes dead.

There is an obvious and relatively common exception. The one thing you do see people with radiculopathy lose completely is active ankle dorsiflexion. The L4 and L5 nerve roots do share the work of dorsiflexion, but in some people (there's a lot of inter-individual variation (78)) L4 does so little that an L5 radiculopathy can cause a foot drop.

It is also worth noting that this is another difference between 'normal' radiculopathy and cauda equina syndrome. In cauda equina syndrome, multiple nerve roots are injured, so loss of bowel, bladder and sexual function can be total.

9

HOW MECHANICAL PRESSURE CAUSES RADICULAR PAIN

'When a telephone cable is cut, the phone falls silent. Damaged nerves behave differently'

— MARSHALL DEVOR (79)

You probably have an intuitive sense that injuring a nerve is very painful. But if you stop to think about it, it is not obvious why this should be. After all, the job of a sensory nerve is to feel; if there is no nerve, shouldn't there be no feeling? In fact, there often *is* no feeling in the skin! How can there be no feeling but at the same time pain?

Adding pressure to healthy nerves does not cause radiating nerve pain. True, if you gently palpate your ulnar nerve at your elbow (your 'funny bone'), you will find that the nerve is sensitive. That sensitivity is mediated by the 'nerves of the nerve' or the nervi nervorum (80). It is probably faint and hard to describe, unpleasant but not quite painful. But it stays in the elbow; it doesn't radiate, like radicular pain. Even if you *bang* your ulnar nerve accidentally, the pain is much more severe but it's still localised to the elbow. (At most you might also feel a brief shower of pins and needles in the little finger side of your hand. This is probably because the sudden force on your ulnar nerve disturbs the resting potential of its neurons and causes a flutter of action potentials to travel up and down the nerve, some of which make their way to the brain and feel like pins and needles in the fingers.)

Nerve roots are no different. The surgeon Stephen Kuslich performed a study in which he stimulated the tissues of awake patients during spinal decompression surgery (58). He found that a healthy nerve root is 'completely insensitive to pain. It could be handled and retracted without anaesthetic. Forceful retraction over an extended period of time resulted in mild paraesthesias but never any significant pain'. In their experiments, Smyth and Wright found that if they pulled a nylon thread tied around a healthy nerve root, it did produce slight radiating pain, but that 'the nerve root which had been pressed upon by a disc was much more sensitive to the touch of the thread than a neighbouring nerve root which was not involved in the herniation' (20).

If a nerve root is apparently so insensitive to pressure, how does pressure cause radicular pain? As we saw when we looked at how pressure can cause a conduction block, the crucial added ingredient is time. With time, inflammation can build up, oxygen levels can decline, and something called 'ectopic impulse generating sites' can develop (they're also sometimes called 'abnormal impulse generating sites'). All of those three things can be painful. We are going to leave inflammation for now and come back to it later; let's look at the other two first.

A lack of oxygen can be painful

'Pain is the prayer of a nerve for healthy blood'

— ROMBERG

A lack of oxygen hurts. It causes the burning muscle pain at the end of an exercise session. It also causes angina, the pain of ischaemic heart disease. A lack of oxygen causes some of the pain of malignant tumours, because malignant tumours damage local circulation and use up all the remaining oxygen so that nearby tissues have none. A lack of oxygen also causes much of the pain and tingling of carpal tunnel syndrome: patients shake out their hands to restore their circulation and their pain eases. People with radicular pain cannot 'shake out' their pain. But some do benefit from gentle movement. And many report immediate pain relief after decompression surgery, with the degree of relief correlating with the increase in blood flow rate (81,82). So it is likely that part of the clinical picture of radicular pain is caused by a lack of oxygen, too.

How does a lack of oxygen cause pain? Without oxygen, the environment inside and around a tissue becomes acidic, replete with hydrogen ions. Sensing these excess hydrogen ions, ion channels such as acid sensing ion channels (ASICs) initiate action potentials.

A distinct but related cause of pain might be the venous congestion itself. If you have ever seen a patient with a deep vein thrombosis or varicose veins, you will know this. Venous pain is aching and heavy. Again, it takes some time to develop. Veins can withstand three times their normal dilation without pain (83). But with prolonged engorgement of the veins, the cells in their inner lining go rogue and release too many enzymes, which degrade the lining and trigger inflammation.

Pressure can lead to ectopic impulse generating sites

Demyelination and axon degeneration are not, in themselves, painful. But in time, they can allow the development of one of the foundational elements of neuropathic pain: ectopic impulse generating sites. Let's look at how.

As we saw in our earlier tour, in a healthy neuron, ion channels are produced in the cell body and taken along the supply line to the wall of the neuron or to neuron's nerve endings in the skin, muscles and joints. In the nerve endings, these channels work together to start electrical action potentials which then zip all the way to the spinal cord or brain. These action potentials tell us what is going on in our bodies and in the outside world.

But in an unhealthy nerve, these channels do not always go where they are supposed to. Instead, they accumulate in demyelinated patches, mistaking these patches for nodes of Ranvier. They also accumulate in the ends of degenerated axons, mistaking these degenerated ends for nerve endings (84). Once accumulated, these channels are capable of triggering an action potential. The sites where such new action potentials form are called 'ectopic impulse generating sites'. (Ectopic simply means 'in the wrong place'.)

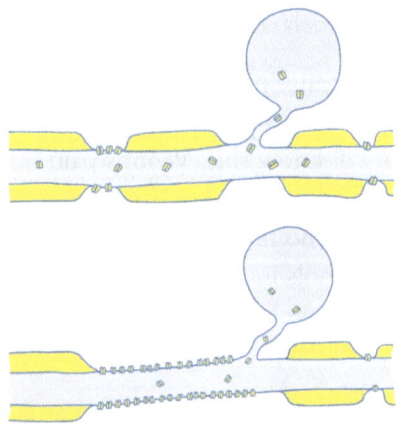

In a healthy myelinated nerve (top picture), ion channels are transported from the ganglion to the nerve endings and, pictured here above, the nodes of Ranvier. In an unhealthy nerve (bottom picture), patches of the axon demyelinate and channels may insert in the bare walls of the axon to create ectopic impulse generating sites.

Ectopic impulse generating sites produce impulses that burst erratically both up and down the nerve. When the impulses that are going up the nerve arrive in the central nervous system, which is so used to receiving, calibrating and processing a steady and predictable stream of sense-information, they stand out like crash cymbals in a violin concerto.

As such, ectopic impulses from nociceptors are often felt as sharp shocks of severe pain (85). About a third of people with radiculopathy have this kind of shooting pain; the same severe symptoms that make trigeminal neuralgia so miserable (86). To make things worse, such an ectopic barrage can 'wind up' the spinal cord into a state of central sensitisation.

As well as causing sudden shocks of pain, some ectopic impulses fire tonically to cause the less severe background pain that is also often a feature of radicular pain. Ectopic impulses are also likely to be behind many of the non-pain sensations that accompany radicular pain, like prickling and pins and needles (we will look more closely at these later).

Although impulses from ectopic impulse generating sites are sometimes described as 'spontaneous', they usually have a trigger. For people with radicular pain, the most common trigger is leg, back and even neck movement, which tensions or compresses nerve roots and mechanically stimulates the channels that comprise the ectopic site. An experiment by Goodwin and colleagues found that the greater the pressure on an ectopic impulse generating site, the stronger the ectopic burst (87). In animal models of radicular pain, particularly strong ectopic bursts from nerve roots have been observed lasting for minutes on end (88). This might be why some patients' pain continues for a short time after a straight leg raise test or spinal movement, for example.

As well as by movement, ectopic impulse generating sites can also be triggered by things like local inflammation, ischaemia, low blood sugars and even circulating stress hormones (79). This last example is one plausible explanation for why people with radicular pain sometimes describe an almost dose-response relationship between stress and pain.

The role of prolonged added pressure in radicular pain is paradoxical. It makes nerves both less active and more active. Nerves are less active because action potentials from the leg fizzle out as they reach a conduction block in the nerve root. And nerves are more active because ectopic action potentials result from demyelinated and degenerated patches of the nerve root.

Again, one of the key points here is that these processes often take a little time - days or weeks, perhaps longer - to build up. The same goes for some of the ways chemical irritation can cause pain, which we'll look at soon. Perhaps this partly explains why some people 'feel something go' in their back but don't feel any radicular pain until a few days or weeks later.

The dorsal root ganglion under mechanical pressure

You might have noticed that, so far in this section, we have been referring to the nerve root but not the dorsal root ganglion. That's because, when it comes to the role of pressure in causing pain, the dorsal root ganglion is a bit different.

Whereas nerve roots only hurt after prolonged added pressure, the dorsal root ganglion can hurt immediately (75,89). An early animal experiment by Howe, Loeser and Calvin found that 'long periods of repetitive firing follow minimal acute compression of the normal dorsal root ganglion' (88). The authors were particularly struck by how long this firing lasted, in some cases for minutes on end. But while the ganglion is sensitive to pressure, it is probably not exquisitely so, particularly in a healthy state. In Kuslich's experiments with awake, locally anaesthetised people, he found that 'the ganglion was somewhat more tender than other parts of the [already sensitised] nerve root, but the difference was not dramatic' (58).

Nevertheless, the ganglion does seem to be at least somewhat more sensitive to pressure than the nerve root. And, like the impulses from ectopic impulse generating sites, impulses from the ganglion travel both up and down the nerve, arriving in the central nervous system as an unexpected burst.

Luckily, the ganglion is not commonly subjected to pressure from a disc herniation. It is usually located out of the 'firing line' of most disc herniations. But in rare cases, the ganglion can be contacted by far lateral herniations and by extrusions that migrate upwards or downwards in the lumbar canal. And it's also vulnerable to stenotic pressure when intervertebral foramina narrow with age. As such, in cadaver studies the ganglion is often found to be deformed or indented, particularly those that happen to be located more proximally (17). As one writer said, the ganglion is 'a very special organ that resides in a very dangerous place' (90). That said, many people with squashed ganglia have no symptoms because foraminal stenosis has such a slow onset and involves relatively little inflammation compared to many disc herniations (17).

Key points on how mechanical pressure causes pain:

- Given time, mechanical pressure can cause pain in the following ways:

1. Pressure reduces oxygen supply to the tissues which can be painful.
2. Pressure causes tissue damage such as demyelination and axon degeneration. At demyelinated sites and degenerated axon tips, ectopic impulse generating sites may develop.

3. The tissue damage can also cause an inflammatory reaction (discussed in the next chapter).

- The dorsal root ganglion is more directly sensitive to mechanical pressure than the nerve root.

10

HOW CHEMICAL IRRITATION CAUSES
RADICULAR PAIN

'*The pain in sciatica cannot be ascribed solely to mechanical pressure: there must be also a pathological alteration of one or more nerve roots, which results in hyperalgesia.*'

— LINDAHL, 1966 (91)

We saw earlier how chemical irritation, independent of any pressure, can be sufficient to cause a conduction block in the nerve root complex. It is similar with pain. In fact, some studies find that mere contact from a disc herniation causes no less pain than full compression, implying that chemical irritation has a substantial role to play in the clinical picture (92).

How exactly does chemical irritation cause pain? Broadly speaking, there are two ways: inflammation directly makes neurons fire, and inflammation sets up ectopic impulse generating sites.

Inflammation 'tells' neurons to fire and become hyperexcitable

Earlier, we saw that when immune cells sense danger they rush to the nerve root, pass through the blood-nerve barrier and clear up and break down myelin sheaths and axons. Once they are in the nerve root, immune cells release *inflammatory mediators*.

Inflammatory mediators are the immune cells' little helpers. They break down tissue, get everything set up so the tissue can regenerate later, and communicate between immune cells to coordinate the whole project. Here's a few inflammatory mediators, which you don't have to remember: IL1β, IL-6, NGF, TNFα, bradykinin...

As you will know if you have ever tried to get your head round this, it's confusing. Each mediator does a few different jobs, and each job is done by a few different mediators (93). And while some mediators trying to do one thing, some other mediators are trying to do the complete opposite thing. Luckily, according to the researcher Ru Rong Ji, despite all this complexity, inflammatory mediators 'modulate pain in surprisingly consistent ways' (94). Ji says they 'either promote or dampen pain depending on the specific identity of the mediators involved'. In other words, if there are more mediators that say to a nociceptor 'Fire!' than there are mediators saying 'Don't fire!' then the nociceptor will fire. And that's how inflammation sensitises neurons and makes them fire.

There might be a temptation to think of these inflammatory mediators as nasty invaders that suddenly force neurons to fire. From one perspective, that's true. But it's also important to see that immune cells and inflammatory mediators are always talking to neurons, and also that neurons are always talking back. They are conversation partners. This 'cross talk' conversation helps to tweak inflammation levels so the body can regulate itself (95,96). The sensitisation that we see after a disc herniation is part of this conversation.

There's also a temptation to think of all this inflammation as a kind of useless curse of life. From one perspective, that's true, too! But it's also important to see that acute inflammation is a normal response to anything that the immune system thinks might be dangerous. It's not necessarily pathological. As Louis Gifford wrote, 'a damaged nerve, in the process of trying to recover and rewire itself, may in some people produce pain [...] That pain is an unfortunate by-product of a very smart regenerative process' (9).

Shouting 'Fire!' to sensory neurons is just one way that some inflammatory mediators cause pain (97). Other inflammatory mediators cause pain indirectly by telling nociceptors to express more ion channels, including voltage gated sodium channels (95). As we have seen, those are the channels that allow nociceptors to fire in response to the world around it. So, the more channels a nociceptor expresses, the more excitable it becomes. With enough extra channels, nociceptors depolarise more easily, more often and sometimes spontaneously!

Inflammation can set up ectopic impulse generating sites

We have already seen that ectopic impulse generating sites can develop where a neuron is demyelinated or its axon degenerated - i.e., where it is damaged. But ectopic impulse generating sites can also develop in the absence of neuron damage, merely in response to inflammation.

How does this happen? Recall that ion channels such as sodium channels are transported along a neuron. If these channels reach a stretch of the axon that is inflamed, they cannot pass and their journey is interrupted (99). Unable to proceed, they cluster in place and embed in the membrane of the axon. After a few days, enough channels are embedded in the axon to create an ectopic impulse generating site (100).

These inflammation-induced hot spots are mechanically sensitive, and the stronger the mechanical stimulus, the more strongly they will fire (87). The researchers who have led the discovery of this mechanism, Dilley and Bove, believe these hot spots give off a tonic (i.e. ongoing) discharge that is felt as a dull ache, and, when provoked, a sudden volley of discharges that is felt as a sharp pain (101). 'The axon membrane at the treatment site', they write, 'behaves as a remote extension of the peripheral terminal'. When inflammation dies down, the bunched-up channels disperse and the site stops generating ectopic impulses.

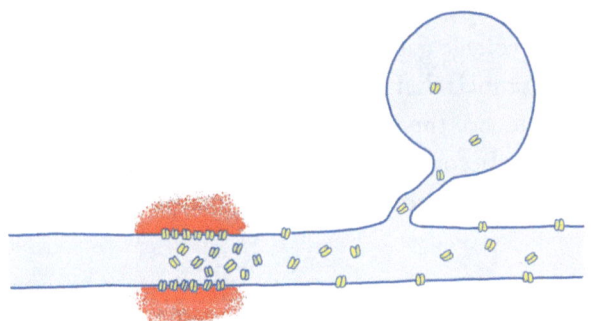

Inflammation can interrupt axonal transport and cause an ectopic impulse generating site

Dilley and Bove's work is important because it shows that neurons can develop ectopic impulse generating sites without being truly damaged. The neurons in their studies do not undergo a major breakdown of their blood-nerve barrier and subsequent infiltration of destructive immune cells, and there is hardly any of the axon and myelin degeneration that we saw earlier. As such, they continue to conduct impulses normally (87). These 'hot spots without (much) damage' could well explain why many patients have radicular pain with no apparent loss of function.

The dorsal root ganglion in inflammation

Let's give some more attention to the dorsal root ganglion. This is a good place to do it because the ganglion is particularly vulnerable to chemical irritation.

Devor called the dorsal root ganglion 'an odder beast than most of us realise' (75). Recall that the ganglion houses all the cell bodies of the primary sensory neurons of its spinal level. So, it is pretty valuable. But for some reason, unlike the cell bodies of the lower motor neurons, the ganglion is located outside the fortress of the central nervous system. Not only that, but it also has a thin capsule and no blood-nerve barrier to protect it. And it is jumpy: as we have seen, it is unique in its ability to spike its own action potentials. Why is the dorsal root ganglion so odd?

The prevailing theory, first put forward by Devor in 1999, is that the ganglion is exposed so that it can sense circulating toxins and inflammatory chemicals (75). And it is jumpy so that it can tell the brain about those toxins and chemicals. That is, the ganglion is a chemoreceptor. There are other chemoreceptors in the body, most notably the part of the brain that senses toxins in the blood and triggers the vomit reflex. In the case of the ganglion, it might be what causes your limbs to ache when you have the flu. This ache is not nice, but it is helpful in the long run because it stops you from expending too much energy moving about while you are sick. If Devor's theory is true (unfortunately it is hard to prove or disprove evolutionary hypotheses), maybe we can think of the ganglion as being like a scout sent out from an army's camp to surveil the area: although it's vulnerable, it can gather important information.

The dorsal root ganglion's possible role as a chemoreceptor also makes it particularly jumpy if it is chemically irritated by a disc herniation. Like the axons of the root, when disc material is applied next to the dorsal root ganglia of rats, within a few hours venous congestion and edema build up (68). Macrophages inside release inflammatory mediators (102). The whole nerve root complex might then become spontaneously active and more sensitive to movement (103). In animal models, the impulses it sends off reach and excite the thalamus (104,105). Not only this, but in response to chemical irritation the ganglion also produces more ion channels such as voltage gated sodium channels, which will make the whole neuron more sensitive (106,107).

Given its role as a chemoreceptor, the dorsal root ganglion is probably sensitive to all kinds of chemical irritation, not just that caused by disc material. Chemotherapy, for example, can chemically irritate and sensitise dorsal root ganglia, which partly explains why people who have undergone chemo are more susceptible to radicular pain, and peripheral neuropathy in general (108). Circulating stress hormones, too, are almost certainly sensed by the ganglion, which might explain why some people with radicular pain often find their pain gets worse with stress. This cross-talk between the stress system and the dorsal root ganglia might also be mediated by structural changes: it's been shown in animal models that, after a nerve injury, sympathetic nerves actually infiltrate the ganglion, forming a basket-like formation inside it (109).

While the ganglion can be jumpy in response to direct mechanical pressure and chemical irritation, it also responds to any injury at any portion of the nerve, including the nerve root (110). It is not clear exactly how the ganglion knows another part of the neuron is injured, but it is not particularly surprising that it would, given that a sensory neuron is all one cell. In animal models, when disc material is applied to the nerve root but *not* the ganglion, the ganglion nevertheless increases its expression of sodium channels (those ectopic impulse generators) (98), experiences a reduction in blood flow, and even initiates programmed cell death.

How important is the dorsal root ganglion for radicular pain? Although the ganglion is relatively out of the way of disc herniations, the fact that it is so jumpy, even to nerve injuries some distance from it, means it surely has some role to play in most cases. It fits some aspects of the picture. For example, although ectopic impulses from the nerve root can sustain themselves for some time, prolonged discharge is more characteristic of the ganglion (88), and this might explain why radicular pain can go on for some time after it has been evoked. Also, as we have seen, many people with chronic radiculopathy have axonal dieback, which might imply an injury to the ganglion. There are not many studies in humans that could tell us for sure how much or how little the ganglion has to do with the picture of radicular pain. In one of the few studies of this kind, Aota and colleagues found that ganglion swelling and indentation of the ganglion correlates with the degree of leg pain (111). In another, North and colleagues found that people with radicular pain do have spontaneous ectopic activity in the cell bodies in their ganglia, whereas people with axial lower back pain alone did not (112). However, the participants in this study had a particularly severe cause of their radicular pain, metastatic spinal cord compression. It may therefore be difficult to generalise to other causes of radicular pain.

Do annular fissures leak disc material?

Before we finish looking at chemical irritation of the nerve root complex, it is worth looking briefly at a theory that says that inner disc material need not herniate to irritate a nerve root, but can instead leak through small fissures in the outer layer of the disc, which is called the annulus. Some researchers believe this accounts for many cases of nerve root pain without any herniation on MRI (113).

Animal models of such an injury do cause nerve root damage (114). And one study, by Peng and colleagues, has found that many people with radicular pain but no herniation did have annular tears on their

MRI, and that the location of these tears correlated with the location of pain (115). The tears seemed to cause slightly different symptoms to herniations. Rarely did sufferers have pain on straight leg raise, or sudden pain provoked by movement of the lumbar spine. Instead, they had a consistent pain and numbness, irrespective of how they moved. The study by Peng and colleagues paints a compelling picture of a subtype of non-compressive, inflammatory radicular pain. However, at present the theory of annular tears remains a mostly theoretical mechanism that can't be identified reliably in clinic.

Key points on how chemical irritation causes pain:

- Broadly speaking, inflammation causes nerve root pain in two ways:

  1. Immune cells release inflammatory mediators that directly sensitise nociceptive primary sensory neurons, and can even 'tell' them to fire.
  2. Inflammation makes travelling channels bunch up in the nerve and form an ectopic impulse generating site. This can happen with little or no structural damage.

- The dorsal root ganglion is particularly receptive to inflammatory mediators, possibly because picking up on them is part of its day to day job.
- The ganglion can become inflamed when a nerve root is injured, even if the ganglion itself is not directly affected.
- The ganglion and, once sensitised, the nerve root are probably both receptive to stress hormones.

11

THE ROLE OF THE CONNECTIVE TISSUE OF THE NERVE IN RADICULAR PAIN

As we've seen, the nerve root sleeve is continuous with the dural sac. Both are primarily made up of dura mater. They are somatic tissues, and probably have some role to play in the somatic referred pain that often accompanies radicular pain.

For example, when Smyth and Wright looped their nylon threads through the dura mater of the dural sac and pulled, it caused an aching pain in the back or the buttock (20). Here's how they describe one such case:

> '*On pulling the dural thread, an ache was felt at the medial and inferior aspect of the left buttock just at the fold. It was localised to an area about one inch in diameter and was of a mild nature.*'

This fits with Cyriax's observation that dural referred pain can sometimes be peculiarly localised (116). Maybe dural referred pain explains why some people feel a deep, one-sided buttock pain for a while before they develop 'full blown' radicular pain. Dural pain might also explain so-called piriformis syndrome and gluteal trigger points (and their upper-limb equivalent of peri-scapular pain).

Discussing the nerve connective tissue is also a good time to mention the *nervi nervorum*, which are the 'nerves of the nerve' that reside in the dural sac, the root sleeve and indeed throughout the connective tissue of the peripheral nerves, too (117). Many of these nervi nervorum are nociceptors. In an inflammatory environment, like any nociceptors, they can become sensitised. As such, they initiate nociceptive pain of neural origin (strange, isn't it?) and can also induce somatic referred pain.

Sensitised nervi nervorum are likely to be responsible for the aching pain some people feel when a nerve is tensioned or pressed, sometimes called neural mechanosensitivity (118,119). For example, you might have seen patients with sciatica who have a pulling, aching pain down their leg and a lot of neural 'tightness', but no severe, sudden or burning ectopic radicular pain. Neural mechanosensitivity, also common in other peripheral nerve conditions like carpal tunnel syndrome, is most plausibly caused by nervi nervorum sensitivity.

That said, it's important not to overstate the role of connective tissue in radicular pain. To go back to Smyth and Wright's experiments on the dural sac, for example, in some patients a thread through the dural sac caused no sensation at all. As Smyth and Wright said,

> *'[Dura mater pain] contrasted strongly with the sharp darting pain caused when the loop around the nerve root was pulled on. The dura mater appeared to be virtually insensitive, the nerve root hypersensitive. The dural thread was pulled a second and third time, but the patient was scarcely aware that anything was being done... [The pain] could not be defined precisely and might be described more accurately as a mild form of discomfort'* (20).

So the dural sac and root sleeve are probably only inconsistently part of the picture of radicular pain. And it's important to restate that neither dural referred pain nor nerve trunk mechanosensitivity are, strictly speaking, *radicular*, since they don't involve ectopic impulses.

But they're often an accompanying part of the picture. Their role is one reason the phrase 'nerve root complex' is a useful one. It helps us to remember how many different kinds of tissue might be contributing to the pain experience.

Key points on the role of the connective tissue in radicular pain

- The nerve root sleeve and the dural sac can occasionally cause referred buttock and leg pain, which is often quite localised.
- The connective tissue of the nerve root and the peripheral nerves in the leg are all innervated by nervi nervorum. When sensitised, these nervi nervorum might be the cause of pulling, aching nerve mechanosensitivity.

12

THE ROLE OF CHANGES IN THE CENTRAL NERVOUS SYSTEM

The spinal cord and the brain react to sensory impulses, which includes the ectopic impulses sent up by nerve root complex injury. They can dampen down incoming sensory impulses, or they can amplify them. (In fact, they do both at the same time, but on balance more one than the other). In the brain, incoming impulses from an injured nerve root complex are integrated with all the other incoming impulses from the rest of the body and its environment, along with an individual's past experiences and present beliefs and expectations. Depending on how this process goes, the impulses from the injured nerve root complex may or may not be transmogrified into the experience of... pain!

That much we know. What we do not know much about, particularly with regard to radicular pain (which has rarely been studied outside the nerve root complex), is the see-saw problem of how far the pain experience is maintained by the central nervous system and how far the pain experience is maintained by the nerve root injury.

It's beyond the scope of this book to discuss the question and, frankly, well beyond the capability of its authors to settle it! But, in response to a pervasive error, it is worth saying that *there is little to no evidence at present for an autonomous central mechanism that can drive nerve pain alone, without any sensory input* (120). Anaesthetic nerve blocks almost always temporarily abolish even chronic peripheral nerve pain (120,121), which implies that at least some afferent information, however little, is needed as a drumbeat for the central nervous system's dance. Any idea that chronic radicular pain has somehow shunted to the central nervous system and is being maintained there might eventually be proven correct, but at the time of writing it is pure ideology.

That said, let's look more closely at what we do know about central changes in radicular pain.

For a start, there is a good deal of evidence that central changes do occur. In both animals and humans, nerve root injuries increase sensitivity in areas of the body outside the affected nerve's territory. And that's not only in extra-territorial parts of the affected limb. For example, people with chronic radicular pain have a lower pain threshold on the same side of their face (74). People and animals with radicular pain in one leg also have an increased sensitivity in the uninjured leg, a kind of mirror image (122,123). For most people, this is almost always subclinical, although it might account for some instances of bilateral sciatica.

We often see extra-territorial sciatica, that expands beyond the expected narrow dermatome in which we are told we will find it (8). This too implies central changes, such as a widening of the receptive field in the sensory nervous system.

However, it must be said that there are some explanations for extra-territorial radicular pain that have nothing to do with central changes. For example, some people have cross-links between their nerve roots, or even conjoined roots that exit together through the same foramen (124). This would cause incoming ectopic impulses to switch and scramble between nerve territories before they even reach the spinal cord. And even normal, un-conjoined nerve roots do not always innervate the expected territory. One might spread its afferents out across more space in the leg than expected. On top of that, much of the extra-territorial aspect of extra-territorial sciatica might just be somatic referred pain, confusing the picture.

Nevertheless, central changes likely account for some of the extra-territorial spread of some people's radicular pain. And certainly they account for those changes in the opposite leg and face! In this book, we're not going to stop and look at central sensitisation, which has been covered extensively elsewhere (125) and is not particularly unique to radicular pain. It will be more useful to the reader if we spend some time looking at *neuroinflammation*.

Neuroinflammation of the spinal cord

Earlier, we saw that when a nerve is exposed to chemical irritants, it becomes inflamed. In one sense, neuroinflammation is simple: as its name implies, it's just 'inflammation of the nervous system'. But neuroinflammation is also surprising because it does not just flare up at the site of injury but also spreads throughout the nervous system.

By 'spread', we do not mean that neuroinflammation expands out like water through pipes. Rather, it flares up in various important processing sites, sort of like a riot in one city might inspire another riot in a neighbouring city. We have looked at this happening within a sensory neuron, when an injury to a nerve root can cause inflammation to flare up in that nerve's own processing site, the dorsal root ganglion. But, particularly if the injury is pronounced, inflammation might also flare up in the dorsal and ventral horn of the spinal cord, in the thalamus and in the brain too (126). These processing sites receive and combine incoming neurons from all over the body, which means that when they are inflamed they can cause a wide territory to become sensitive. This partly explains why some people with radicular pain have widespread or even contralateral pain.

Here's an example. A study by Albrecht and colleagues demonstrated the spread of neuroinflammation in people with lower lumbar radicular pain (127). The investigators used a sophisticated imaging technique that shows neuroinflammation as 'lit up' patches on a scan. They found patches of neuroinflammation at the site of root injury in the foramen, but also in the spinal cord (they did not look in the brain). Considering the lumbar spinal cord ends at about L1 junction, that is some distance for neuroinflammation from the lower lumbar foramen to spread! In a similar study they found that patients with radicular pain had more of these patches of neuroinflammation in some areas of the brain compared to patients with lower back pain only (128).

Let's look more closely at how neuroinflammation in the spinal cord words. It is principally co-ordinated by glial cells (129). Glial cells have their fingers on the neuroinflammation thermostat and are always on the lookout for reasons to increase or decrease the level of local inflammation. They increase inflammation in response to a peripheral nerve injury, like a nerve root complex injury.

The first type of glial cell to respond are the *microglia*. Each one is an autonomous cell which monitors its own domain in the central nervous system. Their octopus legs are constantly surveilling their environment, looking for threats. At the same time, microglia support the neurons in their domain, tweaking their synapses, keeping them healthy and clearing away any debris.

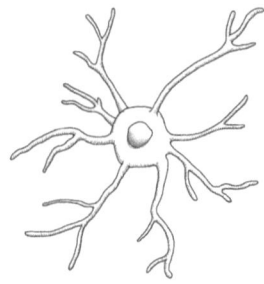

Microglia

When microglia respond to an injury, in this case a nerve root complex injury, they become 'activated' (although that word is a bit misleading since they are always doing something). When activated, microglia pull in their octopus arms and increase their cell body size. They multiply and move to the site of danger and other related parts of the spinal cord and brain. The activated microglia now express a swarm of inflammatory mediators similar to the one we saw in the nerve root. At first, the swarm is heavily populated by pain-promoting ones so, on balance, it will promote pain (94).

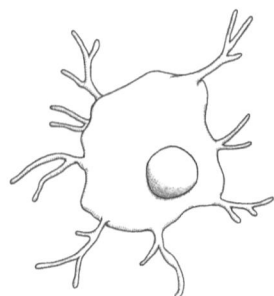

Activated microglia

This frenzy of microglia activation is usually short-lived. But while they are activated, microglia signal to another kind of glia, *astrocytes*. Astrocytes, as their name suggests, look like stars.

Unlike microglia, they are connected to each other in a fine, fibrous network. This network enwraps synapses (as many as one million synapses per astrocyte) and connects with blood vessels. They constantly communicate with neurons, checking that everything is okay. When activated microglia signal to this network of astrocytes, the astrocytes also become activated.

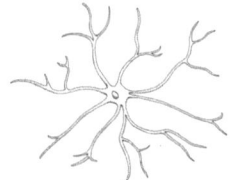

Astrocyte

When activated, astrocytes tell neurons to produce more inflammatory mediators, which adds to the swarm released by microglia, and also to produce more neurotransmitters, which join a feedback loop to further sensitise neurons (notice again that a neuron is not just a passive listener, but an active responder). Activated astrocytes also tell the blood-spinal cord barrier to become more permeable, just as the blood-nerve barrier did in the nerve root complex. This allows immune cells such as T-cells to enter the usually-protected spinal cord (130). There, these immune cells can release still more pro-inflammatory mediators to add to the others, and they can also damage neurons to the point of demyelination and axon degeneration.

All this becomes a feed-forward mechanism in which glia and T-cells make neurons hyperexcitable, and hyperexcitable neurons in turn further activate glia and T-cells. You might have seen images of rodents' spinal cords after experimentally-induced neuroinflammation (131). The injury-side of the cord is packed with activated glia. This glial cell activation might explain why, after a first bout of radicular pain has calmed down, even the slightest thing can sometimes trigger a relapse: animal studies show that after the pain of

A T-cell, an example of an immune cell

a nerve root injury has passed, glia remain activated for some time so that subsequent injuries cause more pain than expected (123,132).

Neuroinflammation is normal and expected, but becomes a problem when it goes on too long

Like inflammation anywhere, neuroinflammation in the spinal cord is an essential protective response to danger such as tissue damage and pathogens. Immune cells clear up debris and fix damage following injury. Degeneration precedes regeneration. And inflammatory mediators work up nociceptors into a state of excitability so that the ensuing pain makes us rest temporarily while the immune system does its work to support tissue healing. In other words, acute neuroinflammation, and the pain associated with it, is a useful protective process.

Neuroinflammation becomes pathological when it goes on for too long. The people who took part in Albrecht's imaging study had had pain for years on end!

Why would this happen? One simple explanation is that the original injury of the nerve root may not have fully resolved, especially the root is still being compressed. As such, continuing irritation of the nerve root may maintain low grade neuroinflammation over a long period of time.

Another explanation is that the neuroinflammatory process itself fails to resolve. One might expect the neuroinflammation process we saw above to just fizzle out after a while. But this isn't how neuroinflammation is resolved. Instead, neuroinflammation should be resolved *actively*. Microglia, immune cells and nociceptors should gradually 'switch sides' and release pain-dampening anti-inflammatory mediators (126). Those anti-inflammatory mediators should build up in the soup of pro-inflammatory mediators, until they outnumber and overwhelm them, making everything go quiet. It could be that a failure of this active resolution process explains many cases of chronic neuroinflammation, such as the people who took part in Albrecht's study (127).

(Incidentally, the fact that inflammation does not fizzle out passively but instead resolves itself actively helps to explain why interfering with an acute inflammatory response can unexpectedly *increase* the risk of developing persistent pain (71). In interfering with inflammation, we are trying to stop a moving train, but we might also be taking away its brakes).

Differences in neuroinflammatory response partly explain why people with similar injuries have very different levels of pain and suffering

As you likely know, two people with apparently similar disc herniations can have very different levels of pain. As we'll see in Part 3, there are many reasons for this - but one is that different people might have different neuroinflammatory responses. The reactivity and readiness of our immune systems are very different; just think about how some people develop bad hay fever in spring, whereas others are not bothered at all by high levels of pollen.

There are many things that influence the reactivity and readiness of our neuroinflammatory reactions. One example is sex differences: neuroinflammation differs between male and females (133). For instance, females rely more on their adaptive immune system than males. This reliance on adaptive immunity might explain why women have greater immune responses following infection or vaccination and therefore often better protection. However, it also partly explains why females are at higher risk of developing auto-immune diseases. Another sex difference is that oestrogen promotes the release of pro-inflammatory mediators from immune cells, whereas testosterone increases production of anti-inflammatory mediators by macrophages. Differences like these partly explain why women are more likely to have neuropathic pain (134) and seek care for sciatica (135).

Another thing that influences the reactivity and readiness of our neuroinflammatory response, and therefore the pain we feel, is our gut microbiome (136). In a setup called the gut-microbiota-brain axis, the gut and the brain 'talk' to each other using hormones (for instance, cortisol), immune cells and proteins (for instance cytokines) and neural signals (along the vagus nerve and the enteric nervous system) (137). This axis affects the central nervous system and appears to help induce and maintain central sensitisation (138).

The gut-microbiota-brain axis has also been used to explain a link between inflammation, pain and low mood. For example, an abnormal composition of gut microbiota in mice with painful peripheral nerve injury are associated with anhedonia (a reduced ability to experience pleasure) and depression-like behaviour (139). And the low mood and pain improved if mice were given probiotics to normalise the gut microbiome. It is therefore not surprising that the theory that inflammation and mood are closely linked has gained some traction in recent years (140). Even though much remains to be explored in this area, we know that nerve injury, neuropathic pain and low mood often co-exist (141). The microbiota-gut-brain axis might be one explanation for this interaction.

The over-arching point here is that neuroinflammatory changes in the nerve root and central nervous system are linked to multiple systems in the body. This partly explains why apparently similar disc herniations can cause drastically different levels of pain and suffering in different people.

All that said, we don't want to get carried away with the idea that neuroinflammation is the key to explaining all that's unexplained about sciatica. We still don't know how much of the picture it explains, or how often, because we cannot easily look at immune cells in the nervous system in humans. The methods we do have for studying neuroinflammation in people with sciatica suggest that its role is inconsistent. For example, when researchers look for inflammatory markers in the intervertebral discs of people with radicular pain, they find those markers in some people but not others (142). When they look in the blood, they find not much evidence of raised inflammatory markers at all (143,144). This fits with what we see clinically, which is that anti-inflammatory treatment like steroid injections work wonders for some people but does little for others.

Key points on the role of central changes in radicular pain:

- There is no evidence that persistent radicular pain is primarily driven by a central mechanism.
- Changes to the central nervous system almost certainly contribute to the clinical picture of radicular pain, most notably in the spread of symptoms beyond the expected dermatomal distribution.
- Neuroinflammation is inflammation within the nervous system. In some cases, an injury to a peripheral nerve can cause neuroinflammation in the spinal cord and brain.
- Neuroinflammation in the spinal cord is similar to the inflammatory processes we saw earlier, but with glial cells playing a dominant role.
- Some people think that nerve pain becomes persistent when the normal resolution processes of neuroinflammation (both central and peripheral) fail.
- Evidence of systemic inflammation in people with radicular pain is inconsistent.
- Differences in immune responses in individual people may account for differences in pain experiences

- Neuroinflammation is complex and important beyond pain. It may be modulated by the gut microbiota and explain low mood associated with nerve injuries. But we do not understand enough about these links as yet.

13

THE ROLE OF THE NERVE DISTAL TO THE NERVE ROOT COMPLEX IN RADICULAR PAIN

By definition, radicular pain is caused by an injury to the nerve root complex. But does the distal part of the nerve, travelling out of the spine, down the leg and into the muscles, joints and skin, have any part to play?

Certainly, the distal nerve is probably not necessary for radicular pain. The nerve root complex is sufficient. This is proven by the fact that people who have had their legs amputated can get phantom radicular pain.

Phantom radicular pain was first described by the physician Arthur King in 1956 (145). One of his patients was a machinist whose leg had been amputated below the knee after an accident at work three years before. This machinist developed 'full-blown sciatica' after he slipped off a wagon and twisted his back. His pain radiated from his back, down the back of his leg and curled, in King's words, 'across the dorsum of the absent foot and into the great and second toes, which were not there.' Imaging showed that the machinist had a disc herniation. After an operation to remove the herniation, his pain got better. The machinist's only lingering problem was some tingling in the absent toes.

Another of King's patients was a labourer whose leg had been amputated above the knee after a gunshot wound became gangrenous during World War I. The labourer developed sciatica after lifting a bucket. The pain, in the patient's words, 'shot out of my back, down my butt, out of the stump and into my big toe that hadn't been there for years.' Again, imaging showed a disc herniation. King operated on this patient too, and his pain improved.

Since King's descriptions, there have been more case reports of phantom radicular pain (146), including reports that corticosteroid injections can relieve it (147).

That said, there probably is often a role for the rest of the neuron peripheral to the nerve root complex in radicular pain. After all, the neuron is one long cell, monitored and regulated by one cell body, so it would be surprising if an injury to one short, proximal part of it didn't lead to more distal changes that also contribute to the picture. And indeed, as we've already seen, radiculopathy is often accompanied by neuron dieback, which possibly contributes to some of the numbness we see clinically.

As another example, after a peripheral nerve injury the dorsal root ganglion produces more ion channels, including sodium channels, which, as we have seen, help to start off an action potential. This increased production can partly be seen as a failure of normal regulation due to injury, but it's also in part a 'deliberate' defensive strategy (148). Either way, a nerve injury means the ganglion produces more channels, many of which embed themselves across the length of the axon's membrane as far as the sensory tips in the tissues. All these extra channels in a neuron means a lower threshold for action potentials. That means that many injured neurons not only have ectopic impulse generating sites at the site of injury, but also channel-stuffed tips that are hyperexcitable to external stimuli, too. Although this is a well-established consequence of peripheral nerve injury in general, ion channel changes to peripheral nerve terminals have to our knowledge not been demonstrated yet in people or even animals with radicular pain. Nevertheless, we know that some patients with radicular pain are hypersensitive to heat stimuli applied to the skin in the affected limb, and such thermal hyperalgesia is thought to be a clinical correlate of peripheral sensitisation (149) as may happen with channel-stuffed tips.

Key points on the role of the nerve distal to the root in radicular pain:

- The fact that people can have phantom radicular pain suggests that the distal axon is not necessary for the feeling of radicular pain.
- That said, an injured neuron is likely to become hypersensitive generally, and this probably contributes to the clinical picture of radicular pain.

14

SIDE NOTE: NEUROPATHIC PAIN

What is neuropathic pain?

According to the International Association for the Study of Pain, neuropathic pain is pain 'caused by a lesion or disease of the somatosensory system' (150). This makes it distinct from the other two mechanistic descriptors of pain according to IASP, which are nociceptive and nociplastic pain (151). Patients with neuropathic pain might feel burning or cold-like pain and electric shocks. They also often report unpleasant sensations like tingling and pins and needles or itch. Some people may additionally have mis-sensations called dysesthesia; for example, a woolen feeling or the feeling of having bubble wrap around the leg. Classic examples of neuropathic pain are post-herpetic neuralgia, caused by a virus that affects the nerves, and painful diabetic neuropathy, caused by metabolic damage to the peripheral nerves. Evidently, neuropathic pain is quite distinct from nociceptive pain.

So, is radicular pain neuropathic? Well, as we've seen, radicular pain is caused by ectopic action potentials emanating from the nerve root (4), which is generally accepted to be a cardinal feature of neuropathic pain (64) that, most people would say, qualifies as 'a lesion or disease of the somatosensory system'. From that perspective, radicular pain is neuropathic. And when the radicular pain is accompanied by radiculopathy, there is still more evidence of a neurological lesion or disease.

However, this perspective is somewhat theoretical - in clinical practice, as we've seen, that ectopic, neuropathic radicular pain is often accompanied by a jumble of non-neuropathic causes of leg pain, such as neural 'mechanosensitivity', somatic referred pain and centrally-mediated or 'nociplastic' pain. Clinically, then, how can we identify neuropathic pain in our patients with spinal leg pain?

Diagnosing neuropathic pain

You might have heard of screening questionnaires such as the pain-DETECT or DN4. They focus on patients' pain descriptors, looking for the characteristic symptoms that suggest pain is neuropathic - tingling, prickling, burning, and so on. These questionnaires can distinguish between patients with neuropathic pain and nociceptive pain pretty well, but not perfectly. They usually miss around 10-20% of patients who we would clinically diagnose to have neuropathic pain (152). That means that neuropathic pain questionnaires can be a helpful tool, but do not replace a careful clinical examination.

Another tool is the clinical grading system (153) developed and revised by The Neuropathic Pain Special Interest Group of the International Association for the Study of Pain (abbreviated to NeuPSIG). This grading system uses a patient's history (including pain descriptors and pain behaviour) and pain distribution, sensory signs, and confirmatory diagnostic tests to determine the likelihood they have neuropathic pain as 'possible', 'probable' or 'definite'. These three categories highlight the challenge of detecting neuropathic pain: it is not always clear cut, and clinicians and patients will have to live with some uncertainty.

A new version of the NeuPSIG grading system, reproduced at the end of this chapter, has recently been developed specifically for use with patients with spine-related-leg pain (287). Let's go through it step by step.

1. Interview. Let's imagine your patient has the symptoms of neuropathic leg pain - perhaps a burning pain with tingling and prickling in a roughly dermatomal distribution. With this information, you can classify them as having 'possible' neuropathic pain.

2. Examination. Next, you can examine your patient to see if you can be more certain of your diagnosis. Let's say they have no apparent loss of strength or sensory function. In this case, there's no 'objective' evidence of loss of nerve function; in other words, there's no evidence of a 'lesion or disease' of the nerve. Therefore, we have to limit our neuropathic pain diagnosis to 'possible'. Many patients with radicular pain will fall into this category, because strength and sensation testing are often normal. On the other hand, let's say that when you examine your they do in fact have objective evidence of a loss of nerve function, for example a loss of sensory function in roughly the same dermatome as the pain. Now, you have more evidence of a nerve lesion. So you can upgrade your neuropathic pain diagnosis to 'probable'.

3. Confirmatory tests. Finally, let's say you also have MRI evidence that your patient's nerve root is compressed. Now, you have a confirmatory diagnostic test so you can upgrade your certainty once again to 'definite' neuropathic pain. The authors of this system note that realistically, you often won't be able to (or won't want to) get an MRI of your patient's spine, in which case a diagnosis of 'possible' neuropathic pain is sufficient to make decisions about medication and so on, considering the rest of the clinical picture.

Like most overarching classification systems, this might feel a little stilted. But it is a reasonable way to navigate the uncertainty around neuropathic pain and clinical practice in general.

Does it matter?

Well, if you are helping your patient to decide whether or not to take anti-neuropathic pain medications, it certainly matters whether their pain is neuropathic. Guidelines clearly state this (154). Far too many patients take anti-neuropathic medications for nociceptive pain, including many patients with referred or mechanosensitive 'sciatica'. So, the distinction is important when it comes to prescribing.

It's also important from a screening perspective. People with neuropathic pain have lower quality of life, higher depression and anxiety and use more healthcare resources than patients with nociceptive pain. Identifying neuropathic pain alerts us to specifically screen for these things. Patients often only open up about the impact of their radicular pain on their emotional wellbeing or social interactions when specifically asked about it. Of course, one could reasonably argue that this kind of screening should be an integral part of our assessment no matter what type of pain patients have. But, unfortunately, it often gets lost in a busy and time-pressured healthcare setting.

But perhaps the most valuable reason to understand and identify neuropathic pain is that doing so can help many patients to make sense of their story. We know that people with radicular pain find the suddenness, severity and strangeness of their neuropathic symptoms to be frightening (155). After all, neuropathic pain is very distinct from the nociceptive pain most people are used to. Not only are patients frightened by neuropathic symptoms, but some even doubt whether those symptoms are real, wondering if it's 'all in the mind' (156). For many people, the concept of neuropathic pain makes the confusion become clear. It acts as a kind of Rosetta stone, translating their cryptic symptoms into a language they can understand. On top of that, the concept of neuropathic pain can also help patients to communicate with their family and friends, find and connect with other people like them (there are neuropathic pain support groups, for example), and navigate the healthcare system.

In the spirit of understanding neuropathic pain, in the next chapter we will look more closely at what causes the extra 'nervy' symptoms that confuse and concern people with radicular pain.

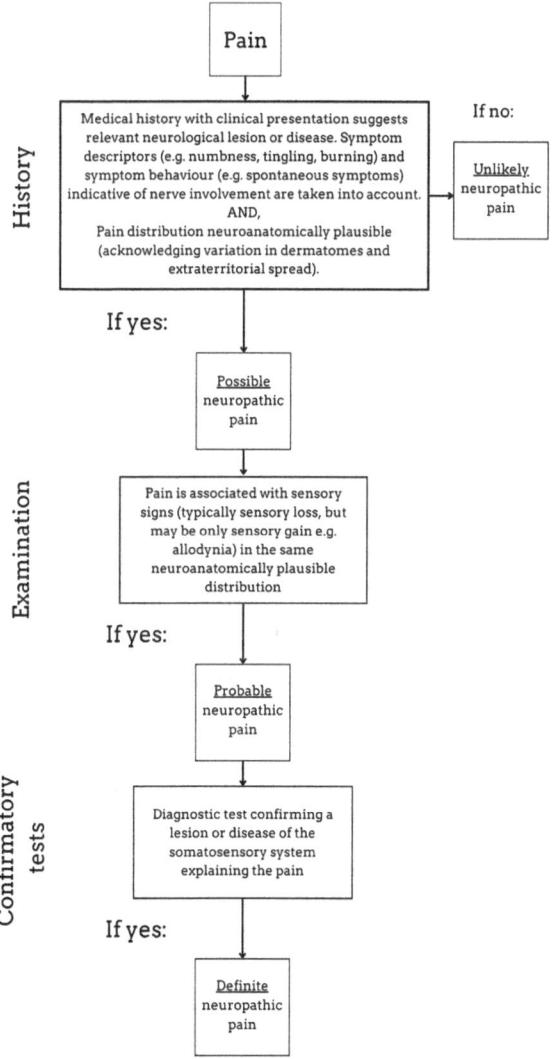

Clinical grading system for the identification of neuropathic pain in people with spine-related leg pain. Developed by a working group of The Neuropathic Pain Special Interest Group of the International Association of Pain (287).

15

SIDE NOTE: WHAT CAUSES THE EXTRA 'NERVY' SENSATIONS

Although nociceptive pain is unpleasant, it is usually predictable. If you twist your ankle, it hurts to walk on it; if your shoulder joint aches, it hurts to move it. And most people have substantial experience of nociceptive pain, too. Everyone has had a paper-cut or stubbed their toe or twisted their back. This means that when people have nociceptive pain, they can usually make some sense of it.

The same cannot be said for neuropathic pain. Whereas neuropathic pain can have a mechanical component, neuropathic pain often comes and goes unpredictably, and co-occurs with a whole host of unusual sensations. In fact, for many people with radicular pain, unusual, non-painful sensations are more prominent than the pain. One study found that 40% of people with painful radiculopathy said that things like prickling and burning were their main symptoms, and not the sudden, shooting pain we classically associate with the condition (86). And these unusual sensations can be just as bothersome as the pain (157).

What causes the unusual sensations? Some, for example the sensation of having a heavy leg or a leg that's wrapped in cotton wool, are probably caused by a loss of sensation and/or strength. A loss of sensation can also make the leg 'feel numb' (which is a kind of paradox - if something's numb, how can you feel it? But consider that if you've ever had your jaw anaesthetised for a dental procedure, you don't feel as if you have no jaw, you 'feel' a numb jaw... in fact, but your numb jaw might also feel strangely swollen, another sensation that is also often described by patients with nerve injuries).

However, most of the unusual sensations of radicular pain are probably caused not by the loss of function but by the simultaneous gain of nerve function; i.e., by ectopic impulses and hyperexcitability.

As you likely know, we have different types of sensory afferent neurons - C, $A\delta$ and $A\beta$ fibres - which carry sensations from different stimuli. It's the type of neuron affected by a nerve injury, along with the pattern of discharge, that determines the neuropathic pain sensation (85). For example, burning pain is probably caused by a gain of function in C fibres, which normally convery heat nociception (and the pain of eating a chilli) (86). And sensations like pins and needles, pricking and tingling are probably caused by a gain of function in $A\beta$ fibres (158), which normally convey light touch information. More complex sensations are probably caused by a combination of different aberrant signals. For example, the sense of water running down the leg, surprisingly common for people with radicular pain, is probably caused by impulses from movement receptors and cold receptors (159). The complex firing patterns of neurons reminds the brain of 'wetness' sensations and thus creates this dysaesthetic sensation.

The particular kind of sensation that ectopic impulses cause also partly depends on the pattern of the discharge - its rhythm, its speed, its distribution. For example, let's say someone with neuropathic pain feels a crawling sensation on their skin. That's probably caused by a gain of nerve function that is roughly similar in rhythm, speed and distribution to the normal sensory neuron activity that would be caused by actual ants crawling on their skin. The difference is that for someone with neuropathic pain, the impulses are coming from ectopic sites in the axon and not, as they normally would, from the sensory endings in the tips of the neurons in the skin.

Of course, what kind of sensation is caused by ectopic impulses or neuronal hyperexcitability ultimately depends on the spinal cord and the brain. As we've seen, the spinal cord too can become hyperexcitable in response to a nerve root injury, expanding its receptive field and responding more strongly to input from the periphery (160). And the brain must do the work of (mis)interpreting all this input.

Unpredictable pain

Why can these unusual sensations, and radicular pain itself, come and go so unpredictably? Some of the unpredictability of radicular pain is because ectopic impulse generating sites have strange patterns of electrical behaviour (mostly depending on how many sodium or potassium channels they have) (79). At some sites, the membrane potential quavers on the threshold of depolarisation, ready to be tipped by the slightest stimulus into an action potential. At others, the membrane potential sits beyond the threshold and cycles through a constant state of spontaneous, rhythmic firing. At still other sites, the membrane potential straddles the threshold and depolarises with a rhythm that is bursty and arrhythmic. It is clear how these strange electrical patterns could produce sensations that come and go unpredictably.

The unpredictability of radicular pain and its unusual sensations is also partly explained by the fact that ectopic impulse generating sites fire not only in response to movement and pressure, but in response to a variety of triggers. Depending on which transductive elements become embedded in a site, they might fire in response to ischaemia, high or low blood glucose, inflammatory mediators or mechanical stimuli, for example (79). Some can also fire in response to circulating stress hormones. This might explain why some people find that as they become more stressed, their sciatica gets worse.

And finally, some ectopic impulse generating sites simply discharge spontaneously.

Cold legs

Let's look more closely at some specific sensations that are common for people with sciatica. Firstly, cold. Some people with sciatica wear a thick woollen sock even in the summer because their foot always feels cold. One reason for this is that injured nerves can become hyperexcitable in response to the cold. We don't quite know why. Historically, this sensitivity was understood as a consequence of abnormal processing in the spinal cord (161). But more recently it has been found to be a consequence of sensitisation of the peripheral nerve. One likely cause is that injured neurons lose some of their potassium channels, which normally act as brakes on action potentials (i.e. they lower the membrane potential). Another explanation is that injured neurons gain too many cold-sensitive ion channels, for example the TRP channel TRPM8, which is the receptor for cold (and unsurprisingly is also activated by menthol) (162,163).

Another reason is that some legs with sciatica don't just feel cold, they are cold! Hakelius and colleagues were the first to document, in 1969, that the affected legs of people with sciatica are colder than their painless opposites (164). After surgery, when patients' pain had resolved, their legs returned to normal temperature. The coldness seemed to be caused by a disturbance of the circulation to the skin. The researchers also found that if they blocked the peripheral nerve to the leg, the temperature of the leg returned to normal. They concluded from this fact that the coldness must be caused by the sympathetic nervous system. As you probably know, the sympathetic, fight-or-flight nervous system causes blood vessels in the skin to constrict which makes the skin colder. (This is why when you are frightened, your 'blood runs cold'). Hakelius and colleagues suggested that nociception from a nerve root can make the sympathetic nervous system more active in that particular segment of the nervous system. For people with sciatica, ongoing nociception would mean ongoing sympathetic nervous system activity and an ongoing cold leg.

A more recent study, by Park and colleagues, confirmed Hakelius's findings, and the researchers agreed with Hakelius's proposed explanation that the sympathetic nervous system is to blame (165). But they also found that a significant minority of people with radicular pain had legs that were in fact hotter than normal. People with hot legs tended to have more acute pain or pain of a traumatic onset. Park and colleagues speculated that, like people with cold legs, people with hot legs did have increased sympathetic nervous system activity, but that the injury was so fresh or the trauma so significant that their nerve roots could not conduct it to the periphery. Without any background sympathetic activity, blood vessels dilate. This would make the leg hot.

Night pain

People with radicular pain, and neuropathic pain in general, often find their pain gets worse as the day goes on and is particularly bad at night. The few studies that investigate this find that it is a feature of neuropathic pain more generally (166). There are a few theories about why this is. It might be partly explained by the fact that the body's circulating levels of opioids and inflammatory cytokines have a daily rhythm. Spinal microglia seem to have a daily rhythm too. Another reason might be that blood pressure naturally drops at night, which might make pain from lack of oxygen worse (remember the reversal of the pressure gradient we discussed earlier). Whatever the cause, it is a common pattern.

Pins and needles

What about pins and needles? One cause seems to be prolonged pressure on nerve roots, which elicits ectopic impulses from Aβ fibres (and perhaps from sub-populations of C-fibres, too) (158). In his experiments, Kuslich described that a normal nerve root was 'completely insensitive to pain' but that 'forceful retraction over an extended period of time resulted in mild paraesthesias' (58). Smyth and Wright reported something similar in their experiments: 'pulling on the loop [around the nerve root] for fifteen seconds produced a sensation of pins and needles in the toes' (20). And some people with radicular pain can evoke pins and needles by changing position to put pressure on their nerve root (e.g., sitting in a car for prolonged periods of time).

Here's an evocative illustration of how pressure on roots causes ectopic impulses, which are felt as pins and needles. It's from a 1984 study by Nordin and colleagues (167). The authors had a patient with sciatica who was able to induce pins and needles in the lateral aspect of his right foot by straining and flexing his neck. To see what was going on, they recorded the activity of the sensory fibres of his sural nerve while he flexed his neck, bringing on his symptoms. They found bursts of activity in the sural nerve that waxed and waned in tandem with the patient's verbal reports of his pins and needles. Presumably, straining and flexing the neck caused enough stress and stretch on the nerve root complex to induce a shower of ectopic impulses, travelling up and down the nerve root. As the upward shower of impulses arrived in the man's brain it was interpreted as pins and needles. Nordin and colleagues were measuring the downward shower as it arrived in the man's sural nerve.

It is not clear why, exactly, prolonged pressure on nerves causes ectopic activity from Aβ fibres. Presumably, a lack of oxygen has something to do with it. (As you might know, people who have panic attacks and hyperventilate can also get pins and needles from lack of oxygen in the blood). One explanation is that on a microscopic level, both the pressure and a lack of oxygen force a build-up of sodium, hydrogen and potassium ions, and it is this build-up that depolarises the neuron's membrane to cause ectopic impulses (158).

Shooting pain

The phrase 'shooting pain' has emerged as common parlance between clinicians and patients. Clinicians tend to use it to mean a small patch of pain that travels quickly down the leg, like a shooting star. Many patients use it this way too, but some also use it to mean pain that stays in one place and is sudden and severe, like being shot.

Defrin and colleagues looked into what patients with radicular pain are really feeling when they say they have 'shooting pain' (168). Many agreed with what clinicians usually mean, that shooting pain is a quick movement down the leg, like a shooting star. But some patients in fact used 'shooting pain' to describe a relatively slow movement of pain (some said as long as an hour!). Some meant an expansion of pain rather than a small patch moving from one place to another. Some people even felt shooting up rather than down the leg. The use of standardised pain questionnaires has made the language of pain more stereotyped, but apparently neat categories like 'shooting pain' hide a wide variety of patients' experiences.

It is odd that pain should shoot at all. When a sensory neuron is stimulated we feel it in one place (usually in its innervation territory, the receptive field). In their paper, Defrin and colleagues put forward a theory of why radicular pain shoots. They point out that there is a representation of the leg in the dorsal horn, akin to the homunculus of the whole body in the brain. Shooting pain, they suggest, starts when an ectopic burst from the nerve root sends a travelling wave of action potentials across that representation. The travelling wave triggers second-order action potentials to shoot up the spine and, because they arrive in the brain one after the other in the order they were triggered on the somatotopic map, the brain mis-registers that something is moving on the leg. It is a speculative theory. But it does fit with Smyth and Wright's observations that a gentle pressure on an irritated nerve root causes a proximal pain, and greater pressure causes pain to expand distally (20). Gentle pressures would cause a smaller ectopic discharge that, according to Defrin's theory, would not expand as far across the representation of the leg in the dorsal horn.

That said, Defrin's theory that shooting pain is a result of activity in the spinal cord would entail an extensive location of pain, because there are many afferent neurons from different spinal segments in the cord. But, most patients describe a limited line of shooting pain, suggesting that the explanation lies in the nerve root complex after all. For example, perhaps Defrin's travelling wave theory could just be applied to the dorsal root ganglion.

Finally, a different plausible explanation for shooting pain is neuronal cross-talk (169), which is the phenomenon of a strong ectopic impulse from one axon activating neighbouring axons. Like Defrin's theory, this will always be difficult to prove or disprove experimentally in humans.

16

SUMMARY OF PART II

Let's step back to take in everything we've seen so far.

The pain of sciatica is properly called lumbar radicular pain. The loss of nerve function is properly called lumbar radiculopathy. When both are together, it's a painful radiculopathy.

The basic cause of sciatica is an injury to a single nerve root complex (nerve root, nerve root sheath and ganglion) in the lower back. Typically, it will be the L5 or S1 nerve root complex. Although the problem is in the nerve root complex, the brain does not 'know' this and 'mistakes' incoming impulses as coming from the lower limb. Therefore, the symptoms of sciatica are felt in the leg, in the distribution of that nerve root.

The basic mechanisms of injury to a nerve root complex are 1) mechanical pressure and 2) chemical irritation.

Where does mechanical pressure come from?

Typically, a disc herniation. But also, the stenotic changes of ageing, and also any other intra-spinal mass, such as a tumour or an aneurism. Pressure might fully 'squash' a nerve root complex, but it might also 'bowstring' it in its place, distorting the neural and vascular tissue, or merely crowd it out.

How does pressure cause a loss of nerve function, i.e. a radiculopathy?

Pressure stops blood, and therefore oxygen, from getting to the nerve. This means the nerve cannot conduct action potentials, which is called a conduction block.

A relatively small amount of added pressure is enough to cause a reversible conduction block through venous congestion. Venous congestion in turn causes edema, which further increases intraneural pressure, a compartment syndrome in miniature.

A higher degree of added pressure can cause a conduction block by stopping blood from flowing into the nerve, which is called ischaemia. Ischaemia can cause structural damage, in the form of demyelination and axon degeneration. Therefore, ischaemia likely causes a longer-lasting or even permanent conduction block.

In addition to that, the amount of time under pressure and the speed of its onset are also important. More time and greater speed mean a greater injury, as a rule.

So, pressure causes a radiculopathy by reducing blood flow to the nerve, which stops it from conducting action potentials. And, if the pressure is particularly great or applied for a particularly long time, pressure might also cause structural nerve damage.

Next, how does pressure cause radicular pain?

There are three main ways.

Firstly, structural nerve damage induces inflammation, which sensitises neurons.

Secondly, when pressure causes demyelination and axon degeneration, the resulting 'naked sites' of a neuron can become populated with channels and form so-called ectopic impulse generating sites. Ectopic impulses burst erratically into the central nervous system, which can be felt as pain. And, depending on the cadence of their discharge and the type of nerve fibre effected, these ectopic sites are also the most likely cause of the extra, 'nervy' sensations, such as pins and needles.

Thirdly, pressure stops blood from flowing into the nerve. The body's tissues are well-equipped to sense a lack of oxygen, and nociceptors will fire accordingly.

So, although mechanical pressure alone does not typically cause nerve pain, if it is severe or sustained it will cause ischaemic pain, nerve damage with subsequent inflammation and lead to ectopic impulse generating sites.

Next, where does chemical irritation come from?

Inflammation triggered by neuronal damage, which we have just mentioned, is one source of chemical irritation.

Additionally, disc material itself appears to be a source of chemical irritation to a nerve root complex. This is partly because many discs that herniate are already degenerating, and partly because the inner part of the disc is effectively a foreign body to the immune system.

On top of that, pressure is itself a cause of chemical irritation, via structural damage.

And, chemical irritation might often come from nearby structures that aren't the disc, such as injured and inflamed facet joints.

How does chemical irritation cause a loss of nerve function, i.e. a radiculopathy?

When irritated, the nerve's blood-nerve barrier opens, allowing circulating immune cells to travel through. They join resident immune cells that are already inside the nerve and break down nerve tissue. Although it might seem strange, this is part of the immune system's response to threat: degeneration precedes regeneration.

How does chemical irritation cause radicular pain?

As well as breaking down nerve tissue, immune cells also release a soup of inflammatory mediators. Broadly speaking these inflammatory mediators are pro-nociceptive: they signal to neurons to 1) tell them to fire, and 2) tell them to express more channels, which makes them more sensitive to other stimuli.

Additionally, inflammation can interrupt axonal transport. This means that channels that are being transported along the axon 'bunch up' at the site of interruption to create ectopic impulse generating sites.

Lastly, the dorsal root ganglion is particularly jumpy in an inflammatory environment. This is partly because it appears to be a chemoreceptor that constantly monitors and responds to circulating chemical irritants. Even in an uninjured state, the ganglion can start its own action potentials, and probably does so in response to such irritants. On top of that, the ganglion is also the manufacturing centre of the neuron, and in response to irritation it produces new channels which serve to sensitise the whole cell.

Mechanical pressure and chemical irritation are not wholly discrete, but interact

As we mentioned at the beginning, it is useful but somewhat artificial to separate pressure from chemical irritation. It's true that some kinds of sciatica are far more driven by one than the other. Foraminal stenosis for example, is predominantly a cause of pressure, whereas an annular fissure, or the herpes zoster virus, are predominantly causes of chemical irritation. But the most common cause of sciatica, a disc herniation, is likely to cause pain through both pressure *and* chemical irritation.

And, as we saw earlier, mechanical pressure can lead to chemical irritation, and vice versa.

It's probable that when mechanical compression and chemical irritation are combined, they cause more pain than either factor alone (122, 170).

And what about the rest of the neuron beyond the nerve root complex?

First, let's look peripherally. There is evidence that people with radiculopathy have a loss of intra-epidermal nerve fibres. It's not clear whether this is because of neuron die-back or whole-cell death. Either way, it's likely to be behind some of the skin numbness we see in patients with radiculopathy (although a conduction block at the root is probably still the main cause).

It is also very likely that, in response to an injury at the nerve root complex, the whole length of the sensory neuron, including its surviving receptive tips, can become hyperexcitable.

Also, the connective tissue of the nerve probably becomes sensitised after a root injury. The connective tissue is somatic tissue, innervated by 'nervi nervorum', and so the resulting sensation is probably the aching, pulling, nociceptive pain that is sometimes called 'heightened mechanosensitivity'.

Finally, let's look centrally. Central sensitisation is almost certainly very common for people with radicular pain. It's likely to be driven by ectopic impulses, as well as being a physiological reaction to nerve injury.

Additionally, many people with radicular pain probably have central neuroinflammation. Like peripheral nerve inflammation, this involves an opening of the blood-nerve barrier and an infiltration of immune cells, which might break down neural tissue and release pro-nociceptive inflammatory mediators. This central inflammation might 'amplify' peripheral input from the injury to the nerve root complex, as well as widen the nervous system's receptive field so that radicular pain is felt in a wider distribution.

That said, there's no evidence that any central mechanism is the 'driver' of radicular pain. Most likely, central changes are sustained by peripheral input.

We hope that summary was useful. In the second half of this book, we're going to look at everything that's around the nerve root, including the vertebrae, the foramen, the nearby ligaments and, of course, the all-important disc.

PART III: THE DISC

17

THE SPINE AND THE RADICULAR CANAL

We've looked closely at the nerve root complex, and at the great length of the primary sensory nerve beyond it. But, apart from our brief tour at the beginning, we've been looking at the nerve as if it were floating by itself in a glass jar! So now let's look properly at where all this stuff sits in the body, what's around it, and why that matters.

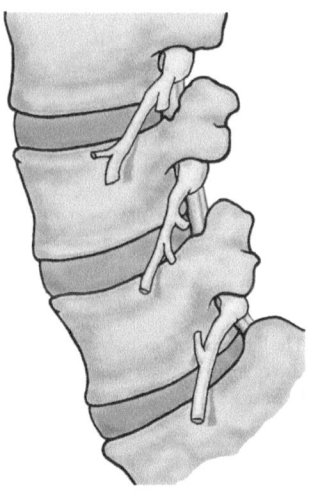

In the lower lumbar spine, as we saw earlier, a nerve root hangs downward in the dural sac and then veers to the side to exit on a slant. From this

Lateral view of the lower lumbar spine

slanted angle, it continues to swoop outwards, as if pulling out of a nosedive, into the intervertebral foramen. It never quite levels out of its nosedive onto the horizontal plane, but by the time it leaves the foramen a root is at about a 45 degree angle from its direct descent through the dural sac (171).

The part of a nerve root that is outside the dural sac but has not yet left the foramen is inside the *radicular canal* (172). The canal is not a literal structure, but rather a space left between other structures so the nerve root can pass through. The radicular canal is longest in the lower lumbar spine because the roots have further to travel from the spinal cord which, as we know, ends in the upper lumbar spine.

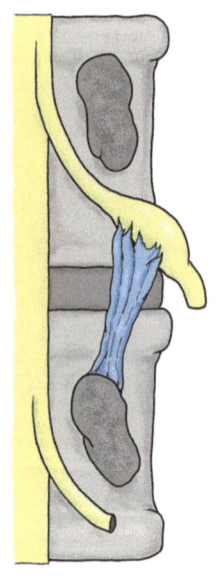

The ligament of Spencer, also called the lateral root ligament

In the radicular canal, the root is exposed to the surrounding structures of the spine. Most notoriously, it is exposed to the disc. You might not have noticed before that the root is held forwards in the canal, against the vertebra and, in places, against the disc. The roots are also relatively fixed in their position. At the proximal side, they are fixed by the tight funnel where the dural sac branches off to become the nerve root sleeve. And at the distal side, they are fixed by a latticework of transforaminal ligaments and, tying the root to the pedicle below it, the ligament of Spencer (173).

This eponymous ligament was first described by Spencer in 1983 (174). Spencer found that when he applied pressure to the nerve roots from the disc side, a lateral ligament prevented the roots from displacing as far as the opposite wall of the canal. By his reckoning, this rendered the size of the disc herniation moot: what mattered was not the herniation per se but how tightly the ligament held the root in place as the herniation bore down. As it turns out, there is a good deal of variation in the strength of the ligament, even between different levels of the same person's spine (175). In fact, some levels have no ligament of Spencer at all. Nevertheless, as a rule the nerve roots are relatively restricted to the disc side of the radicular canal.

The location of a nerve root injury

Lastly, let's look again at why the position of the roots matters when it comes to which ones are more likely to get injured, and by what kind of disc herniation. You might expect the nerve root to be injured by a disc herniation at the level where it exits. It might seem this way because, looking at a diagram at the end of this chapter, this is where we see the greatest length of exposed root. The exiting root just looks vulnerable. But, looking closely, you can see that the exiting root actually hugs the ceiling of the foramen passageway, which means it's out of the way of the disc. And, in any case, disc lesions are rare in this far lateral position.

Instead, the nerve root is most likely to be injured by a disc at the level before it exits the spine, just as it is budding off from the dural sac. You can think of this as the 'shoulder' of the sleeve (22). As you can see, it's here that the lower lumbar nerve roots pass over the discs. And not only that, but disc herniations are also much more common here, too.

So if you picture an L5/S1 disc herniation, it is more likely to affect the transiting S1 nerve root rather than the by-now-out—of-the-way exiting L5 nerve root. This explains why S1 nerve root lesions are common despite the fact that there is no such thing as an S1/S2 disc to injure it where it exits the spine.

You can also see in the picture at the end of this chapter that all that doesn't really apply in the *upper* lumbar spine (176). In the upper lumbar spine, the roots exit the dural sac *after* they have already passed the disc, at the level of the vertebrae. So, any herniation of an upper lumbar disc doesn't hit a transiting nerve root, because it hasn't left the dural sac yet. On top of that, upper lumbar herniations are a bit more rare than lower lumbar herniations anyway. These two facts together partly explain why radiculopathy is more common at lower lumbar levels, affecting the L5 and S1 roots.

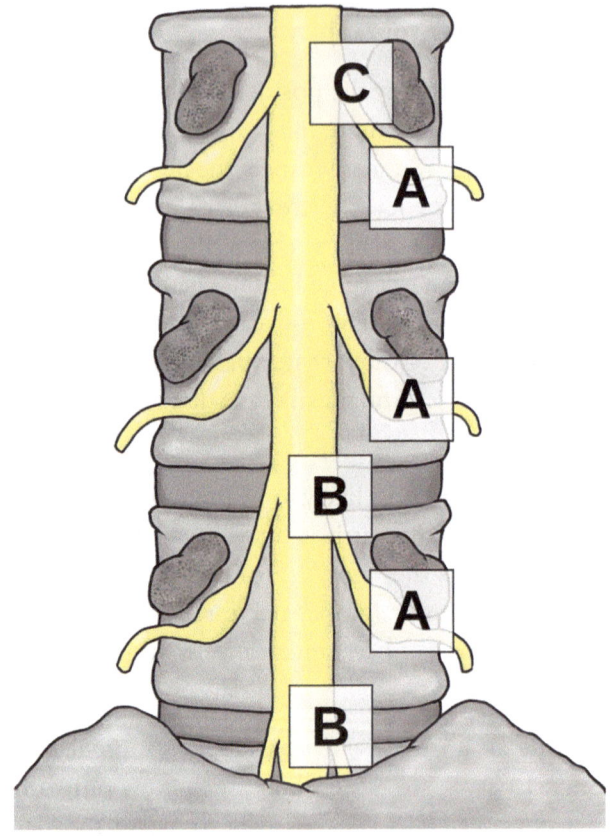

A = exiting nerve root, less affected by disc herniation. B = transiting lower lumbar nerve root, relatively more affected by herniation. C = transiting upper lumbar nerve root, less affected by herniation.

Key points on the spine and the radicular canal:

- Nerve roots sit forwards in the radicular canal, against the disc.
- The ligaments of Spencer holds them in place, although there is a lot of inter- and intra-individual variation.

- Nerve roots are most vulnerable just as they exit the dural sac, because this is where they pass directly over a disc. This is particularly so in the lower lumbar spine.

18

THE LIVING DISC

In 1930, a 25 year old man fell while skiing, twisted his back, and soon after felt a radiating pain down the back of his left thigh into his calf. As Parisien and Ball tell the story (177), the man was referred to the physician Joseph Barr, who prescribed 'several months in absolute recumbency on a Bradford frame', the same restrictive structure used for people with polio.

The man's pain did not improve. In 1932, Barr referred the man to Massachusetts General Hospital, under the care of the surgeon William Mixter. Mixter performed a laminectomy from L2 to S1 and found what he thought was a cartilaginous tumour at the S1 level. At the time, cartilaginous tumours were thought to be the most common spinal cause of sciatica.

After the operation, Mixter told Barr about the tumour. But Barr pointed out that the patient's pain had not developed insidiously as one would expect with a tumour, but had come on suddenly after his skiing trauma. Something didn't make sense.

Mixter and Barr agreed to investigate this discrepancy. Together they reviewed the biopsies from all of Mixter's previous patients with sciatica in whom Mixter thought he had found a cartilaginous tumour. To their surprise, they found that most of the lesions Mixter had excised in fact consisted of disc material.

Before Mixter and Barr's research there had been some speculation that disc herniations cause sciatica (178–181). But Mixter and Barr were the first to investigate this systematically and the first to show that disc herniations are not an uncommon peculiarity but in fact the most common spinal cause of sciatica (182). At first, their research was either ignored or dismissed as too controversial. But gradually it began to take hold and the so-called 'Dynasty of the Disc' began (183).

Looking back on the dynasty of the disc, we can see that the disc became over-blamed for spinal ailments, a misconception that contributed to the overuse of spinal surgery. But we have Mixter and Barr to thank for their discovery that disc herniations are a common cause of radicular pain, and that surgery to take them out often works. Let's look at what we have learned about the disc since 1932.

The appearance of the intervertebral disc

If you want to picture holding a lumbar disc in your hand: it is oval-shaped, sometimes with a faint kidney bean-like indentation on the back wall. It is a finger's breadth in height, thicker towards the front end than the back.

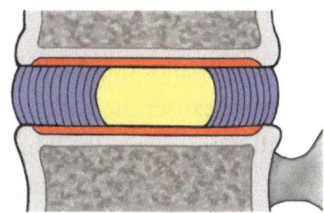

A lumbar intervertebral disc with endplates (orange), annulus (purple) and nucleus (yellow).

The disc is made up of three kinds of tissue. They are distinct, but blend into each other slightly.

Capping off the top and bottom of the disc, tough but also porous, are the *endplates*.

Around the edges is the stiff, tyre-like *annulus*. The annulus is made up of 10 to 20 collagen-rich concentric rings (184), so that if you cut it in half and looked down on it, it would look like a tree stump.

Finally, cloistered in the middle there is the soft and watery *nucleus*. It seems that every researcher has their favoured analogy to describe the consistency of the nucleus. You'll hear that it's like porridge, watery crab meat, heavy phlegm, hair gel, toothpaste... maybe if you mixed all of these things together you would get the best approximation? In any case, the nucleus of a young adult is something between a liquid and a solid, and becomes less liquid and more solid with age.

How the disc deals with force

Simply standing up or walking sends a lot of force axially down the spine. Not only is there the force of body-weight, but also the forces of muscles contracting to keep you upright and balanced. In all, simply standing in place sends about 80 to 100kg of compressive force through the spine (185). How does the disc help the spine cope?

Imagine pressing with your hand on a small block of wood that's sitting on a table. All of the force of your hand is distributed down through the block of wood and into the table. If you touched your fingers to one side of the block of wood, you wouldn't be able to feel any force coming out laterally. This is roughly how force travels through the vertebrae: downward and undispersed.

Now, imagine pressing with your hand on a water balloon, which is also sitting on a table. You can see that the pressure causes the balloon to splay out laterally, and that it does so in all directions equally. This is why the nucleus holds a lot of water: so that, under pressure, it can splay out laterally, and in all directions equally (186).

This doesn't fully solve the problem of how to manage all that force coming down through the spine. The annulus does that. The annulus is a tyre-like restraint around the nucleus. When the nucleus makes its water-balloon motion and splays out at all sides, the annulus accepts that outward force. Now, axial forces coming down the spine are directed not just down but also out, into the waiting arms of the annular rings (186).

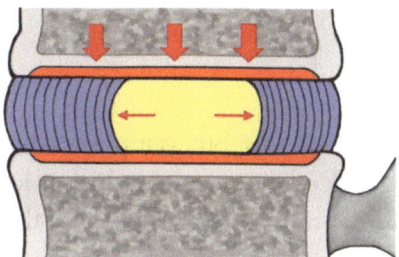

The nucleus, yellow, distributes downward axial force laterally, into the annulus, purple.

The endplates do their part too, deforming slightly when the nucleus pushes into them in a way that the bony vertebrae could not. (However, they are not as strong as the annulus. As such, if downward force is too high, the endplates are usually the first to give in to the outward pressure of the nucleus. The nucleus herniates up or down through them, which creates a Schmorl's node).

In summary, although you might have heard that the nucleus is a shock absorber, that's not quite right. It's better to call the nucleus a *force distributor*.

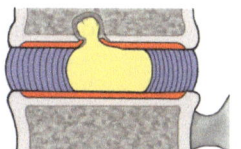

Schmorl's node

Of course, all that is just what happens when we're standing upright. So let's watch what the disc does when we bend. Picture the water balloon again in your mind's eye. This time, imagine pressing your hand down at an angle on one side of the balloon. The side you press on gets compressed, and the other side pushes outward. The same thing happens when we bend down to pick something up: the front of the nucleus is compressed and the back part pushes outward (186), like this:

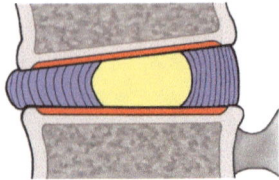

The nucleus can cope with this, but as it pushes outward at the back, it stretches out the back of the annulus, which is less pliable. On top of that, the outward pressure of the nucleus that was previously distributed evenly into the annulus is now disproportionately concentrated backwards. And one more thing: the posterior section of the annulus also happens to be the thinnest and weakest.

Given all this, it is no surprise that in experiments, compressing and flexing animal spines creates posterior herniations with remarkable regularity (187,188). It doesn't happen all at once. Instead, with cycle after cycle of flexion, the annulus gradually loses its integrity. As it does so, the nucleus, repeatedly exerting that outward and backward pressure, begins to nose through the annulus layer by layer. When the annulus loses its integrity from the inner layers first, gradually moving out, it's called a radial fissure:

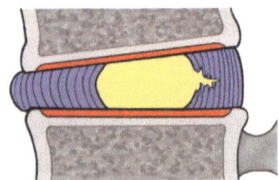

It's likely that this herniation is often forceful, more of an ejaculation than a herniation, especially if the nucleus is very watery and viscous as it is in young people.

Sometimes, a herniation doesn't occur by this classic 'inside to out' pattern, and instead it's the outer layers of the annulus that lose their integrity first (189). This leaves the inner layers with more work than they can manage, allowing a herniation to form. However, whichever part of the annulus weakens first, the effect is the same. Annular defects allow the nucleus to nose its way through and, if the nucleus eventually noses through the last, outer layer of the annulus, it escapes into the 'outside world' as a herniation. In effect, annular defects are the initiation sites for herniations (189).

This gradual-then-all-at-once mechanism for a herniation partly explains why so many people say that it was something like bending down to pick up a sock that caused their back to 'go' and their sciatica to start: that innocuous movement might just have been the last straw for a slowly-developing radial fissure. In fact, most people with radicular pain caused by a disc herniation cannot put their finger on any inciting event, never mind an innocuous one; and those that can identify an inciting event rarely say it was heavy lifting (190).

However, you might be thinking to yourself that there's more to this story than biomechanics - and you'd be right! We'll look at the bigger picture of disc herniations soon.

Movement of the disc

As we have just seen, when we flex the spine the nucleus distributes force backwards. But, somewhat counterintuitively, this does not mean that when we flex the disc bulges backwards; in fact, it bulges forwards. This is because when we flex forwards the vertebrae gap posteriorly, which makes the annulus stretch out at the back rather than bulge. At the same time, the vertebrae close in on one another anteriorly, which makes the annulus bulge out at the front. In extension, the opposite is true: discs bulge out at the back.

This behaviour is not uniform, and it depends on the state of the disc. But, as demonstrated in the upright and dynamic MRIs of people with radicular pain, the fact that the disc bulges posteriorly with extension helps to explain why, despite popular expectations, many people feel more pain in extension and less pain in flexion (191,192). It also partly explains why people with spinal stenosis and neurogenic claudication also feel worse with extension and better with flexion (and like to push shopping trolleys!)

However, although flexion in theory might increase the space for the root, it's not necessarily the case once a disc has herniated. And flexion also pulls nerve roots taut, which can be painful for people whose nerve roots are mechanically sensitive or already significantly irritated. The key message is not that one direction is better than another, but that flexion is not inherently bad and might well bring unexpected symptom relief.

The microscopic disc

You might have noticed that this section opened by promising to show you the living disc but we have ended up talking about experimentally flexed and compressed dead ones. So, let's zoom in to the microscopic level to look at how the disc manages the basic transactions of living: getting oxygen, getting nutrition and expelling waste (193).

Most tissues have blood vessels to do this. In infants, discs do too. But as we grow and start to walk, the pressure inside them becomes too much and blood vessels retreat to the outer few millimetres of the annulus.

To compensate for the disc's lack of blood vessels, the endplate has pores which blend into marrow channels in the vertebral body. These pores allow oxygen and nutrients to diffuse from the marrow, through the endplate and into the disc. The disc gets about 80% of its nutrition in this way (194). (I think the endplate is confusingly named because you don't think of a plate as being porous. But then, 'endfilter' isn't going to catch on).

Nutrients and oxygen don't just wander into the disc, they are lured in by *proteoglycans*. Proteoglycans are the first of three microscopic structures in the disc that we are going to look at. They are spindly filaments of chemicals and proteins that are distributed throughout the disc in complexes of various sizes. According to Bogduk, they are like 'tangles of cotton wool' (186).

Proteoglycans have a negative electrical charge, so they lure in molecules that have a positive electrical charge or are neutral. The nutrients and oxygen in the vertebral bone marrow are neutral, so through the pores of the endplate and into the disc they go.

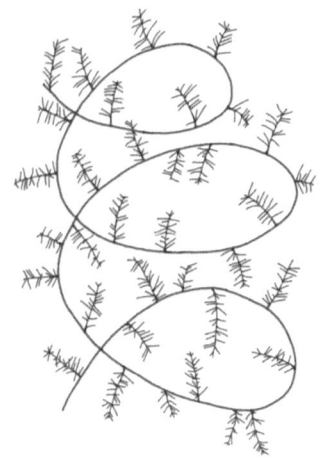

A cotton-wool proteoglycan

The proteoglycans in a disc also lure in another useful neutrally-charged substance: water. As we've seen, water allows discs to keep their nucleus watery so they can distribute force. (You might have heard of this process before if you've ever read up on tendon pathology. In a panic, overloaded tendon cells lay down too much proteoglycan, which lures in water (195). That's why some tendons look swollen. This property of proteoglycans is no good for a tendon, but for a disc it's great.)

Once in the disc, nutrients and oxygen are used for energy by the second of our important microscopic structures of the disc, *chondrocyte cells*. Chondrocyte cells are the craftsmen of the disc. They manufacture the microscopic parts that comprise the whole of the disc, including cotton-wool proteoglycans and entwined cords of collagen.

Unfortunately, chondrocyte cells don't get enough oxygen through the endplate pores to survive in numbers. This means that when chondrocyte cells try to repair any damage done to the annulus by repeated flexion, they find themselves understaffed (194). This means that, despite the efforts of these 'craftsmen' chondrocyte cells, many radial fissures progress unchecked and eventually herniate.

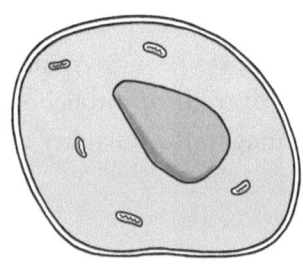

A 'craftsman' chondrocyte

As well as making new parts for the disc, chondrocyte cells also oversee the disposal of old parts. But chondrocytes are the kind of craftsmen who don't like getting their hands dirty, so instead of doing waste disposal themselves they make enzymes to do it for them. These enzymes are called matrix metallopeptidases. Matrix metallopeptidases, which we will just call *MMP*

MMP enzymes

enzymes, are the last of the microscopic elements we are going to look at. MMP enzymes are inactive in their normal state. But they become active when exposed to certain signalling molecules, which chondrocytes partly control. When activated, MMP enzymes break down the cotton-wool proteoglycans and entwined cords of collagen that make up the bulk of the disc, and shoo them out through the pores of the endplates.

You can see how, just like any other tissue in the body, the disc uses oxygen and nutrients as fuel to build itself new parts, and to break down and dispose of old parts as well. The disc has no blood vessels to help, but solves the problem with proteoglycans, which have a negative charge that attracts oxygen and nutrients. To build itself new parts, the disc has craftsmen cells called chondrocytes, although it has fewer of them than it would like. And to break down old parts, these chondrocytes also make enzymes called MMP enzymes.

As in all of the body's tissues, this building-up and breaking-down of the disc is an ongoing cycle that needs all the moving parts - proteoglycans, chondrocytes, MMP enzymes - to be in balance. But you have probably already noticed that this balance is a tenuous one. Next, we are going to see how this balance can be disrupted.

Key points on the living disc:

- The disc is made up of three kinds of tissue:

  1. The endplate, which is porous so that nutrients, oxygen and waste products can pass through.
  2. The annulus, which is tough and made up of concentric rings
  3. The nucleus, which is somewhere between a liquid and a solid.

- The nucleus is a force distributor: when force passes down through the spine, the nucleus disperses some of that force outward, into the rings of the annulus.
- When we bend forwards, the nucleus pushes force back into the posterior aspect of the annulus. For some people, with repeated bending, this causes the annulus to lose its integrity and the nucleus to nose through. This is called a radial fissure.

- If a radial fissure breaks through the outermost layer of the annulus, it becomes a herniation.
- Three important components of the disc are:

1. Proteoglycans, which drag in water.
2. Chondrocyte cells, which regulate the disc and build disc material like proteoglycans and collagen.
3. Matrix metallopeptidases, which are enzymes that break down old disc material.

- Proteoglycans, chondrocytes and MMP enzymes exist in a balance.

19

HOW DISCS GET OLD, GET INJURED, AND HOW THEY DEGENERATE.

The ageing disc

What is ageing, for a disc? As infants, we have small, thin discs that are mostly porous endplate with a little bit of annulus and nucleus. Oxygen and nutrients can easily diffuse in, and waste products can easily diffuse out. As infants we also have a good amount of blood vessels deep into the annulus. That makes two good supply lines.

But over time, as our discs get bigger and thicken, the endplates shrink in comparison to the rest of the disc and, unable to survive the pressure inside the disc, the blood vessels retreat to the periphery. By the time we have had our first growth spurt, it has become difficult for the disc to get enough oxygen and nutrition in and enough waste out (194).

In this harsh, sealed-off micro-environment, the microscopic parts that make up the disc can begin to suffer (185). It happens from the inside out. Proteoglycans, normally aggregated into long clouds of cotton wool, begin to degrade to smaller wisps which don't drag in as much oxygen, nutrients or water. (It is when proteoglycans degrade that the disc appears dark on T2 weighted MRI, because it lacks water. In real life, incidentally, ageing discs are not black but sepia yellow). In a kind of demographic shift, the craftsman chondrocytes that laid down the hydrated nucleus of the young disc gradually die off and are replaced by different cells that prefer to lay down drier annular material. And the MMP enzymes, disturbed by the changing disc environment, multiply and break down more tissue than they should.

Earlier, we saw how the disc lives with an income/expenditure budget that is only just balanced. When proteoglycans, craftsman cells, and MMP enzymes stop working as they should, the disc begins to run on a deficit. That is ageing.

This is not that unusual. All tissues age. But discs do it sooner. This is why epidemiological studies show 'age-related change' on the MRIs of over a third of people in their twenties.

Normal disc ageing is also not painful, just as patches of grey in the hair or crow's feet around the eyes are not painful. And although most old discs do thin out, they keep their basic structure and still do their job. For all the ingenious force distribution of the nucleus, the annulus can hold up pretty well without it. So by the time we are old and our nuclei are dehydrated, the annulus is doing most of the work just fine (196). In fact, there are even some benefits to an ageing disc. Old discs are less likely to herniate because, without big healthy proteoglycans to drag in water, the nucleus loses its ability to exert outward pressure.

But while ageing itself is not painful, it can make the disc vulnerable to injury and degeneration.

The injured disc

We have already seen the two main kinds of disc injury. One, repeated bending can stretch out the posterior annulus - a radial fissure - and force the nucleus slowly back through it until it escapes as a herniation. And two, although we only looked at it briefly, sudden axial load can force the nucleus up or down into the endplate as a Schmorl's node (187).

Of course, that was in cadavers. Like any tissues, living discs like stress. For example, they like the stress of body-weight: Videman and colleagues found that between twins with differing body-weights there was no difference in disc health (197). And discs also like the stress of exercise. The discs of recreational athletes are as healthy, if not healthier, than normal people; the discs of inactive people are less healthy (198–201). By contrast, the discs of astronauts, who experience a prolonged period of *un*loading while they are in space, are less healthy and herniate more often (202). In short, discs like load.

But, like any tissue, there is a limit to how much a disc can adapt. Load that's particularly forceful, particularly frequent or particularly prolonged can test a disc's limits, especially if it is not punctuated by regular rest and recovery (193). This means that, like any other tissues, discs can get injured. However, unlike many tissues, the mostly avascular disc is less able to adapt to injury than many other tissues. This is especially the case when a disc is already ageing, and well into an income/expenditure deficit.

So, discs are strong and adaptable structures, but their lack of a blood supply means that they age relatively quickly, and when they're injured they don't recover as well. However, our day to day experience tells us there is more to it. If it were just about age and loading injuries, then the only people with disc herniations would be middle aged, overworked manual labourers... and perhaps some highly sedentary people who hadn't given their discs enough load for them to adapt! But we know that disc herniations, and radicular pain, can strike anyone in any walk of life.

Genetics might explain this. How much a disc ages is not really a matter of how many years have passed, but what genes have been inherited (203,204). This was first discovered by a number of twin studies from the 1990s (205,206). These studies found that twins who take up very different occupations - farmer and lawyer, plumber and salesman - have, in their middle age, virtually identical spines. These studies have been criticised (207), as there is a danger of over-stating the role of genes in disc health (208). But on the whole, it seems that heredity matters. Some people inherit genes that make their water-dragging proteoglycans more likely to degrade or their destructive MMP enzymes more likely to outwork their craftsmen chondrocytes. And some people inherit genes that do the opposite. In other words, if your patient tells you that sciatica runs in the family, they might be right!

So if a middle aged, overworked manual labourer has been lucky with his genetic inheritance, he might well have young-looking discs. He's less likely to pick up a disc injury and has a better chance of recovering if he does. But, if another middle aged, overworked manual labourer has been *unlucky* with her genetic inheritance, then she might well have had old-looking discs *whatever* occupation she'd chosen.

In summary, 'injury', in the conventional sense, is something that happens to a disc if it is exposed to too much load. But, perhaps more than other parts of the body, a disc's susceptibility to injury and its ability to recover from injury are strongly influenced by its physiological age, i.e. its microscopic health. And its physiological age, in turn, is strongly influenced by both loading history (for both good and bad) and genetics.

The degenerating disc

There is one more piece of the picture. We have seen that the microscopic health of a disc makes it more or less susceptible to, and able to recover from, injury. But there is a relationship in the other direction: injury disturbs the microscopic health of the disc. To return to our financial metaphor: if disc metabolism is a barely-balanced bank account that runs into a deficit as we age, then injury is an unexpected and sudden expense. In attempting to respond to that expense, the disc can slip into a highly disordered pattern of activity that some people call disc degeneration.

Incidentally, the fact that injury is a cause of degeneration explains why some people have one sole degenerated disc, while the rest seem perfectly healthy (something you wouldn't expect to see if disc degeneration was purely genetic). Perhaps that one disc sustained an injury and slipped into a degenerative pattern (209).

How does injury cause disc degeneration? It starts with the disc's craftsmen cells. They are too few and too poorly-nourished to lay down enough collagen to respond to injuries (185). If you could hold a degenerating disc in your hand, you would see clefts in the annulus where the cells have not been able to keep up with demand.

Even while craftsmen cells fall behind on rebuilding, MMP enzymes continue to break down old structures (194). This means waste products build up in the disc. As waste products build up, the pH in the disc, already low, becomes more acidic. In an acidic environment, those cotton-wool proteoglycans break down into thinner and thinner wisps which drag in less and less water. This dehydrates the nucleus, thinning it out. Holding a degenerating disc in your hand, it would be flatter than a healthy disc. If you applied pressure to the disc, the annulus would not only bulge out as expected, but collapse inward too. The dehydrated nucleus can no longer apply the water-balloon-like pressure that would normally push the annulus outward.

In response to stress, the cells of our bodies' tissues end their own lives (210). Like inflammation earlier, this might seem like a self-defeating thing to do. But it allows young and healthy cells to replace them just when we need them most. Craftsmen cells also end their lives in response to the stress of injury and degeneration. But, as you might expect by now, discs aren't good at getting new ones. The already struggling craftsman cell population begins to die off.

We said that ageing discs are not painful. By contrast, although it's difficult to correlate clinically, it's likely that degenerating discs can be. This is because when cells break down their surrounding environment, and as they struggle and die, they release inflammatory mediators (211). These mediators can leak through annular clefts onto the nociceptors in the outer few millimetres of the disc and nearby tissues. Additionally, some people also think that as degenerating discs lose pressure, nociceptors can also grow deeper into the interior to meet the inflammatory irritants half way, although this 'ingrowth' idea is disputed (212).

The build-up of inflammatory mediators from a degenerating disc might explain why some people have back pain for a little while before they get radicular pain. Perhaps the back pain is caused by nociception from the outer annulus of a degenerating disc (it is unlikely to be caused by the nerve root, since stimulation of the nerve root in humans doesn't cause any back pain, only buttock and leg pain).

How they all relate.

As always with these things, the different categories - ageing, injury and degeneration - all overlap and play into one another.

All discs age, some more than others depending on inheritance. Many discs age and age, without ever becoming injured or degenerating. But some discs are so weakened by age that they do injure, or age so much that they tip into degeneration.

Injury and degeneration have a reciprocal relationship. Injury can cause degeneration because the disc cannot keep up with the demands of healing; degeneration can cause injury because it makes the disc weak and more likely to fail with load.

You might have noticed that the difference between ageing and degeneration is not very clear. In fact, according to some people, there is no difference; degeneration, they say, is just ageing but worse. This may be true from a certain perspective. Both ageing and degeneration are caused by the disc running into a deficit. It is just that one is a slight deficit and one is a dramatic deficit that seems to have some runaway effects. But it's true that, like evening slowly becoming night-time, there is no moment you can pinpoint to say that one became the other.

Nevertheless, it helps to look at the two as different. It reminds us that ageing is expected and not pathological. And it helps us to see how injury can knock an ageing disc 'off course' so that it degenerates markedly, while its neighbours remain merely old.

There is another point that people disagree on: how much is a disc herniation a consequence of injury, and how much is it a consequence of genetic inheritance? There have been some pendulum swings in this debate. For a long time after disc herniations were discovered, it was assumed that they were caused by injury. If you work a heavy job, gradually the annulus weakens and the nucleus noses through. Then, after the twin studies in the nineties, a lot of people became sure that genetic inheritance was the thing. If you have the wrong parents and get the wrong ticket in the genetic lottery then the stage is set.

This book is influenced by the work of Michael Adams and Patricia Dolan, which we understand to be a synthesis of the two perspectives (187,209,213). Adams and Dolan acknowledge the importance of genetic inheritance in ageing, but say that often its real role is as tinder for the spark of injury, which in turn causes the fire of degeneration and herniation. We think this interpretation is a good balance. And it helps us to help our patients make sense of things. After all, they often come to us with stories that sound like they are mostly injury, stories that sound like they are mostly genetic inheritance, and everything in between.

Key points on how discs get injured, get old and degenerate:

- Like any other tissue, discs 'like' to move and adapt to load.
- However, unlike many other tissues, a disc is a sealed-off micro-environment with few blood vessels. This means it is hard for oxygen and nutrients to get in, and waste-products to get out. This makes discs more vulnerable than many other tissues to physiological ageing, load-induced injuries and tissue degeneration.
- To expand on these processes.

1. Ageing occurs when the micro-environment of a disc is not getting enough oxygen or nutrients in, or waste products out. Disc ageing is not pathological, but it might make a disc more vulnerable to injury or degeneration.
2. Injuries occur with mechanical overload, typically too much movement with too little rest. Repeated bending can cause a radial fissure, and high axial load can cause a Schmorl's node. Disc injuries are common and often not painful, but can precipitate a herniation.
3. Degeneration is a more disordered and dramatic form of disc ageing. Waste products can build up and the disc environment can become acidic. Disc injuries can trigger or exacerbate disc degeneration because they increase the metabolic demand of a sometimes already-struggling disc.

- Disc ageing, injury and degeneration are all in large part determined by heredity.

20

SIDE NOTE: WHY DO SOME PEOPLE HAVE RADICULAR PAIN BUT NO HERNIATION ON THEIR MRI?

This is very common. There are three categories of explanation for apparently-radicular pain without a herniation:

1. The patient has radicular pain, but the routine clinical MRI is not sensitive enough to see the problem.

Although it is natural to think of the MRI image as like the 'eye of God', able to see everything, this is of course far from the case. MRI images are not very sensitive at all for nerve root problems (214). There are a few reasons for this. One is that many herniations are position dependent, and therefore don't show up on a supine MRI, only appearing in standing (215). Another is that, as we will see, many herniations get smaller and even disappear over time, leaving nothing on the MRI image even though the consequences of the herniation - neuroinflammation, venous congestion - have endured (216, 217). Finally, there are plausible non-herniation causes of radicular pain that would not appear, or not be reported, on an MRI image, for example annular fissures (115), nerve root adhesions (218) foraminal stenosis (219).

An important point to remember is that radicular pain is a clinical diagnosis, not a radiological one. After all, routine MRI do not tell us about intraneural health.

2. The patient actually has somatic referred pain, not radicular pain.

For example, some studies show that people get radicular-like leg pain when their discs are stimulated (220) or lumbar interspinal ligaments are stimulated (3). And sacroiliac joint pain can look very radicular (221).

3. The patient actually has problem with a nerve trunk, not a root.

To name but two examples, deep gluteal syndrome (222) and endometriosis plaques (223) can both injure the sciatic nerve trunk, causing radiating, neuropathic leg pain that has nothing to do with the nerve root complex.

Unfortunately, it's beyond the scope of this book to discuss clinical assessment. An important message, however, is that just because an MRI image or an MRI report says there is no herniation, does not mean that a patient's pain is somehow 'psychological' or 'centrally sensitised'.

21

WHAT'S IN A HERNIATED DISC?

We have already seen how, particularly with ageing and degeneration, the nucleus can nose its way through the annulus to the back of the disc and break through: a herniation.

You might be surprised to hear that herniation material is not necessarily made up of mostly nucleus material. Many herniations also contain fragments of annulus (224). These fragments are presumably broken off and forced backwards into the spinal canal by the nucleus material as it escapes. Annulus is more common in the herniations of older people, because in younger people the annulus is often still too strong to fragment (225).

In fact, in younger people, the outer layer of the annulus is so strong that it doesn't break at all under the pressure of the nucleus (226). Instead, the annulus sometimes peels off a shard of endplate from the bone, and the nucleus herniates through the gap created. Perhaps this isn't too surprising; after all, the annulus is evolved to accept outward force from the nucleus but the endplate is not, and so is only weakly attached to the vertebrae. As such, about a third to a half of herniations contain endplate (226–228) and some studies report numbers that are even higher (229). Many of those fragments of endplate still have a straight edge where they have been peeled off the bone (226). In fact, because annulus also blends partly into the bone as well as the endplate, it sometimes pulls bone off the vertebra too (annulus really is strong!). All in all, many herniations don't seem to actually conform to the classic picture (which we painted earlier) of a herniation nosing its way through the annulus until it finally breaks through the final layer. For many people, it seems like the nucleus does nose through some of the annulus but that the herniation occurs when the annulus, under pressure, wrenches the endplate from the vertebra.

There isn't much data, but people with endplate in their herniations seem to have more loss of nerve function than people who don't. Perhaps a hard shard is more detrimental to a nerve's function than the porridge-like nucleus (225,230). On a routine clinical MRI, endplate material doesn't necessarily show up as different to the rest of the herniation.

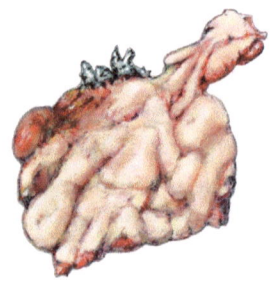

Herniated disc material. Note the red patches of granulated tissue and, at the top, the more jagged edge of peeled endplate and bone.

So a herniation contains nucleus material but also sometimes annulus, endplate and bone, too.

What happens to this herniated material once it's in the world outside the disc? It swells rapidly (185). The cotton-wool proteoglycans, which until now have been luring water into the centre of the disc, suddenly find themselves in a low-pressure environment where they can lure in more water than they had ever dreamed of. A new herniation can double its weight in four hours. Then, this effect reverses as the proteoglycans are degraded and give up all that water.

Over time, however, the mass of the herniation is bulked up again by the tissue that forms around all healing wounds: granulation tissue

(231,232). Granulation tissue is made up of a soft matrix, immune cells and a network of new capillaries that give it a light red colour. It's more likely to form on large herniations that have broken through the posterior longitudinal ligament because they are more exposed to the immune system. Granulation tissue does not seem to have much effect on the clinical picture at first but, as we will see, it does help to resorb the disc, shrinking its mass again in time.

Key points on what's in a herniated disc:

- A disc herniation can contain not only nucleus but annulus, endplate and even bone.
- Contrary to expectation, many herniations breach the endplate or even the vertebrae, rather than the outer layer of the annulus.
- Particularly if it is large and uncontained, a herniation may be covered in granulation tissue. This tissue contains immune cells that can resorb the disc material.

22

TYPES OF HERNIATION

'The attractive black-or-white explanation on the basis of "disc in" or "disc out" has much to commend it, since it has the charm of nice, simple revealed truth and this is always popular, even when descriptions include varieties of 'disc in' and "disc out"'

— GRIEVE (33)

Even if you do not request clinical MRI scans you will have had the experience of reading through a report with your patient to help them to understand it. Let's talk about what the technical words on an MRI report mean so you can be sure of what you are saying (233). First, here's an ordinary disc.

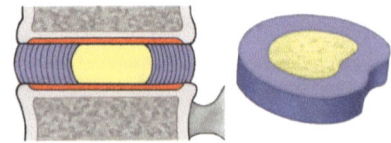

Different shapes of herniation.

It's easiest to start with the difference between a bulge and a herniation. A bulge is a generalised displacement of disc tissue. Sometimes this displacement is around the whole circumference of the disc, sometimes less. But it has to be fairly widespread; by one definition, a bulge must extend across more than 25% (90°) of the circumference of the disc (233).

Although a bulge spreads across a lot of the circumference of a disc, it usually only extends a short distance outward, rarely more than 3mm.

Here's a disc bulge from the side:

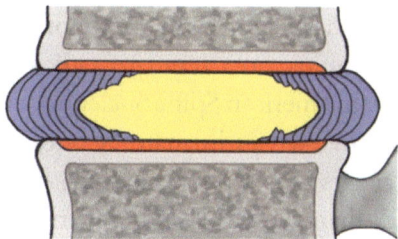

By contrast, a herniation is a focalised or local displacement of disc tissue. By one definition, they take up *less* than 25% (90°) of the circumference of the disc (233). But although they are narrower than bulges, they can extend a greater distance outward. This means that unlike most disc bulges, some large herniations can contact nerve roots.

'Herniation' is a kind of umbrella term. There are two main subtypes of herniation: protrusions and extrusions.

A protrusion is a herniation where the herniation is still contained within the outer layer of the annulus. Here's a protrusion:

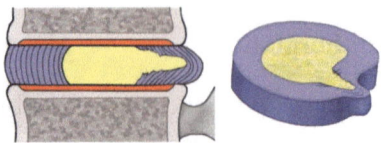

By contrast, an extrusion is a herniation that *has* broken through the outer layer of the annulus. Here's an extrusion:

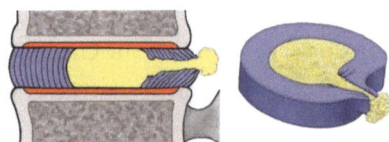

One problem with these definitions is that on imaging, it's quite hard to identify whether a herniation has or has not broken through the outer layer of the annulus. So a standard radiological classification, outlined by the North American Spine Society, doesn't define protrusions and extrusions by their pathoanatomy but by their *shape* (233). This categorisation says that a protrusion is a herniation that has a base that is thicker than the head, and an extrusion is a herniation that has a head that is thicker than the base.

There is also one sub-type of extrusion, the sequestration. Sequestrations are extrusions that are completely detached from the disc. They're sometimes called 'free fragments'. Here's a sequestration:

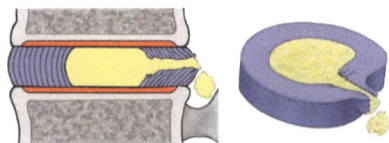

Another important pair of terms are 'contained' and 'uncontained'. Contained herniations exist wholly within the outer layer of the annulus or the posterior longitudinal ligament (the ligament which travels down the back of the discs and vertebrae). Uncontained herniations break through both, into the 'outside world' of the spinal canal.

Perhaps counter-intuitively, uncontained herniations might cause less trouble in the long run than herniations that are contained. This is because uncontained herniations are more exposed to the immune system. They attract more of the granulation tissue that we've just seen, and are therefore more likely to be resorbed.

Different locations of herniation.

Those are the types of herniation, now let's look at the different positions.

First, there are central herniations. These are not that likely to cause radicular symptoms. They butt against the middle of the posterior longitudinal ligament and, behind that, the dural sac. As we have seen, nerve roots are mobile in the dural sac, so they can usually move out of the way of central herniations. When central herniations do cause radicular symptoms, it is often serious. If a central herniation is large enough to compress the nerve roots in the dural sac then this will include the sacral roots, which take care of bladder, bowel and sexual functions. This is cauda equina syndrome.

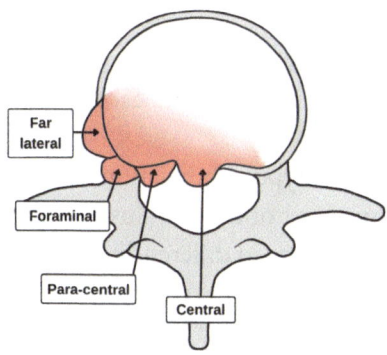

Top down view of the different locations for a herniation.

The most common kind of herniation is paracentral. If central is at the 6 o'clock position, paracentral is 5 o'clock or 7 o'clock. These herniations are most common because the annulus is thinnest and weakest in this position, especially if the disc is more of a kidney-bean shape. What's more, they are often able to break through the posterior longitudinal ligament, or edge around it. As we have seen, it is at the 5 o'clock and 7 o'clock position that the nerve roots of the lower lumbar spine are just budding off from the dural sac. Here, the roots are held like guy ropes, unable to move out of the way of a herniation. Most people with radicular pain have paracentral herniations.

Next, at 4 o'clock or 8 o'clock, are *foraminal* herniations. And anything more lateral than a foraminal herniation is a far lateral herniation. Foraminal and far lateral herniations are rare clinically because the nerve roots hug the top of the foramen as they exit, where they are out of the firing line of the disc. These kinds of herniations tend to become symptomatic when they sequester and travel to contact the root. There, they are said to cause worse pain than paracentral herniations because they contact the jumpy dorsal root ganglion, and perhaps pin the nerve root complex against the roof of the foramen (25,234,235). Far lateral herniations can even contact the spinal nerve.

Key Points on the types of herniation:

- A bulge is a generalised displacement of disc tissue.
- A disc herniation is a focalised or local displacement of disc tissue
- Protrusions are a subtype of herniation that do not break through the outer layer of the annulus. (Radiographically, they are sometimes defined differently, as herniations in which the base is thicker than the head.)
- Extrusions are a subtype of herniation that do break through the outer layer of the annulus. (Radiographically,

they are sometimes defined differently, as herniations in which the head is thicker than the base.)

- Sequestrations are a subtype of extrusion that break free from the rest of the disc entirely.
- Uncontained herniations break through the posterior longitudinal ligament; contained herniations do not.
- Herniations can also be classified by positions: central, paracentral (the most common), foraminal or far lateral.

23

THE ASSOCIATION BETWEEN HERNIATIONS AND PAIN.

Asymptomatic disc herniations are common (except for extrusions).

It's well-known that herniations are common in asymptomatic people (236). However, this seems to be true only of disc protrusions. Disc extrusions are not that common; according to one meta-analysis, they are found in only about 1.8% of asymptomatic people (237). This fits with findings from a study by Suri and colleagues (238), who followed asymptomatic people and intermittently gave them an MRI to see what was going on. Only five of the 123 people got a disc extrusion, and none were asymptomatic - all five extrusions came with radicular pain.

Extrusions aside though, asymptomatic herniations are indeed common. According to one systematic review, 29% of asymptomatic people in their 20s have a disc protrusion, and this rate increases gradually through life, up to 43% of asymptomatic people in their 80s (236). How can a herniation be asymptomatic? It could be that it's not that chemically-active, or that it's developed very slowly, or that it's just not near a nerve root. Of course, it also depends on broader systemic factors that might make someone more or less predisposed to pain, such as their immunological profile.

Sometimes people point to the existence of asymptomatic disc protrusions to argue that herniations don't cause radicular pain. That is hasty generalisation. It would be like saying 'campfires don't cause forest fires, because lots of people have campfires that don't cause forest fires', or 'speeding doesn't cause car accidents, because lots of people speed and don't get in accidents'. X can sometimes cause Y and sometimes not.

Herniations are more common in people with radicular pain.

People who have sciatica are, as you'd expect, more likely to have a disc herniation than people who don't have sciatica. For example:

- In their meta-analysis, Brinjikji et al. (237) concluded that 1.8% of asymptomatic people had a disc extrusion, compared to 7.1% of symptomatic people. And 19% of people had a disc protrusion, compared to 42% of symptomatic people (These people had spinal pain in general - unfortunately no such big review exists for radicular pain specifically.)
- Boos et al.(239) found that 22% of asymptomatic people had herniation-induced nerve root compression (which was mostly minor), compared to 83% of people with radicular pain (which was mostly severe).
- Konstantinou et al. (135) found that 32% of people with somatic referred pain had nerve root compression, mostly caused by disc herniation (some stenosis), compared to 60% of people with radicular pain.
- van Rijn et al. (240) didn't compare symptomatic to asymptomatic people, but the symptomatic to the asymptomatic side of the spine in people with radicular

pain. The asymptomatic side had a herniation 33% of the time, whereas the symptomatic side had a herniation 74% of the time.

The importance of disc herniations diminshes over time

For example,

- El Barzouhi et al. (60) looked at the the MRI images of a group of patients one year after their treatment. They found that about a third of people who got better still had a disc herniation, and... about a third of people who didn't get better also still had a disc herniation. Another way of looking at the data: of all the people who still had a disc herniation a year after treatment, 85% felt better.
- In a similar study, Barth et al. (241) found that two years after surgery, many patients still had herniations (remnants and reherniations), but these findings didn't correlate with their clinical status.
- Fraser et al. (242) found much the same thing, 10 years after treatment - 'the presence or absence of herniation had no significant bearing on a successful outcome'

Essentially, after a year or so, most people's pain seems to be doing its own thing, irrespective of what the disc is doing! It seems that herniations are generally short- to medium-term trigger of radicular pain. Presumably they 'spark' the neural mechanisms of pain such as neuroinflammation, structural nerve damage, ectopic sites, and local scarring and tethering; and, once sparked, those mechanisms are partially self-sustaining.

Bonus (possible) fact: People with leg pain but no back pain might be more likely to have an extrusion than a protrusion

Just an aside based on two small studies:

- Pople and Griffith (243) found that 96% of people with sciatica but no back pain had a disc extrusion.
- Reihani-Kermani (244) found that people with sciatica but no back pain were 6.5x more likely to have an extrusion. And, if their radicular pain increased as their back pain decreased, they were fully 10x more likely.

Next, let's look at herniation size and pain.

Initial herniation size is (probably) not associated with the degree of pain

For example,

- Dunsmuir et al. (245) found that amongst 56 patients 'there [was] no direct correlation between the size or position of the disc prolapse and a patient's symptoms'.
- Karpinnen et al. (246), found that amongst 160 patients 'magnetic resonance imaging [was] unable to distinguish sciatic patients in terms of the severity of their symptoms'. (Interestingly, larger disc herniations were associated with a higher likelihood of having radicular pain, as opposed to sciatica.)
- Mariajoseph et al. (247), amongst 122 patients receiving a microdiscectomy, found that 'disc fragment weight had no effect on the severity of pain'.
- Seo et al. (216) amongst 43 patients undergoing conservative treatment, found that 'no statistically significant correlation was evident between symptom severity and disc volume'.

Not quite the same thing, but related - Jensen et al. (92) found that amongst patients with radiculopathy, nerve root touch caused as much leg pain as nerve root displacement or compression.

It seems that once a herniation is painful, it doesn't much matter if it's big or small.

Of course, this is all just pain. Herniation size does seem to be associated with the degree of neurological deficit (248).

Herniation size is (probably) not associated with the eventual clinical outcome

For example,

- Gupta et al. (249) found that 'there is no statistical association between the size of a lumbar disc herniation and the likelihood that a patient will fail conservative treatment and ultimately require surgery'.
- Masui et al. (250) found that 'Clinical outcome did not depend on the size of herniation' after seven years.
- Modic et al.(251), amongst 246 patients with acute radicular pain or back pain, found that 'there was no relationship between herniation type, size, and behaviour over time with outcome.'

So, the size of the herniation doesn't seem to matter much. That might be because of measurement problems when it comes to MRI (which, after all, is not the eye of God (252,253); and what would upright MRIs have found?(215)). But even accounting for measurement problems, it seems that size doesn't matter much. This fits with what we've seen. Nerve roots are relatively fixed in place on the disc-side of the spinal canal, which means that a herniation doesn't have to be particularly big to put them under significant mechanical pressure. Radicular pain is in large part driven by discogenic chemical irritation, which renders the actual size of the herniation less important. And of course the experience of pain is significantly determined by systemic and psycho-social factors.

The important thing is that a disc herniation is not like a vice that causes pain through mechanical pressure, and causes more pain the more tightly it is wound. Considering the role of this chemical irritation, a closer analogy is that a herniation is like putting chilli powder on your fingertip and poking, or even just touching, your eye. The contact is important, and more pressure is surely not desirable, but the pain is mostly caused by chemical irritation. (Thanks to Rob Goldsmith for this analogy).

Change in herniation size is associated with a change in pain

As we'll see, herniations change in size over time, most of them shrinking as they are resorbed by the immune system, but some growing in size as more disc material escapes. And these changes in size do seem to go along with changes in pain. For example,

- Seo et al. (216) split patients into a group whose herniations got smaller over time and a group whose herniations got bigger. Both groups' leg pain improved on average, but pain improved more quickly in the group whose herniations got smaller.
- Ahn et al. (254) found that a decrease in disc size of more than 20% correlated strongly with a positive outcome. 18 of 19 people whose discs decreased by 20% had relief of most or all of their pain.
- Fagerlund et al. (255) found a 'significant positive correlation between the improvement from sciatic pain and the reduction in the size of the individual hernia', amongst 30 patients

- Kesikburun et al. (256) found that patients whose discs completely resorbed also had less pain, whereas patients with partial or no resorbtion did not, amongst 40 patients
- Komori et al. (257) found that change in herniation size 'mainly corresponded to clinical outcomes but tended to lag behind improvement of leg pain', amongst 77 patients. The lag suggests that the two might not actually be closely related; discs resorb and people get better but the resorbtion might not be the cause of the improvement.

There are exceptions (251), and the association is hardly overwhelming; but it does seem to be the case that a change in disc size is associated with a change in pain.

It seems kind of contradictory that the size of a herniation isn't associated with pain, but change in size over time is. Maybe it's not the shrinking per se that causes a pain reduction, but the fact that a shrinking disc is also resolving as a biochemical event, fizzling out, becoming less inflamed... To put it another way, 'change in herniation size' might be a proxy for 'inflammatory process doing its thing, body working its way back to normal'.

General conclusions

Although asymptomatic herniations are common, asymptomatic extrusions are not that common. And, as expected, people with radicular pain have more herniations than people who don't have radicular pain. That said, the importance of symptomatic herniations decreases over time, because herniations are generally a short- to medium-term trigger for radicular pain.

Once a herniation is painful, it's not necessarily the case that a bigger herniation is more painful, or vice versa. A bigger herniations also doesn't mean a worse outcome in the long run. That said, as herniations get smaller people do seem to feel less pain.

24

HOW DISC MATERIAL CHEMICALLY IRRITATES A NERVE ROOT COMPLEX

Now let's look at how a disc herniation causes pain. Earlier, we mentioned that mere contact from a disc herniation can chemically irritate a nerve root complex. Now, let's look in more detail.

There are two possible explanations for how a disc herniation chemically irritates a nerve root complex. One is that herniations are painful because the interior of the disc, especially if it is degenerating, contains pro-inflammatory chemicals. Let's call this the endogenous explanation, because it points at how pro-inflammatory chemicals are endogenous to the disc. The other is that herniations are painful because the nucleus of the disc is foreign to the immune system, so it causes an autoimmune reaction. Let's call this the autoimmune explanation.

There is a bit of disagreement in the background here. Some people think that it is more endogenous than autoimmune, and some people think the opposite. We are not going to resolve this disagreement, but it's good to know about it!

The endogenous explanation

According to the endogenous explanation, the inside of a degenerating disc contains plenty of substances that could irritate a nerve root (258). For example, as part of their role in breaking down and remodelling the disc, craftsman cells express pro-inflammatory mediators (211). To help them do this, they also release enzymes, including MMP enzymes, which are potential irritants, too. And aside from the stuff they release, craftsmen cells themselves can be destructive. They have been shown in laboratory conditions to phagocytose (gobble up) synthetic micro-beads and dead cells (259).

Not only do the inside layers of a degenerating disc contain plenty of irritating substances, but so do the outside layers (211). As a disc degenerates, waste collagen and proteoglycan build up. In the outside layers of the disc, where there is a blood supply, this waste is detected by blood-borne immune receptors. These immune receptors trigger an inflammatory reaction, just as they would in response to any tissue damage. This means that, according to the endogenous explanation, the outside edge of the degenerating disc is host to an ongoing inflammatory reaction which, when the disc herniates, may irritate the nerve roots. Maybe this explains why some people have grumbling nociceptive back pain for a while before they 'feel something go' and develop radicular pain.

Let's look at some specific examples that support the endogenous explanation. The first, uncovered by Saal and colleagues in 1990, is that there is an enzyme in disc herniations called phospholipase A2, or PLA2 (260). Although you might not have heard of PLA2, it plays a part in lots of stuff you have heard of. For example, it triggers the inflammatory cascade that includes COX1 and COX2, the enzymes that are inhibited by ibuprofen. And it is found in the venom of snakes, insects and spiders. Nasty stuff! Harvesting disc material from five patients undergoing discectomy, Saal and colleagues found 'extraordinarily high levels' of PLA2; hundreds to thousands of times higher than in other tissues of the body. PLA2 activity was higher in non-contained herniations, particularly extrusions, than in contained herniations. It was also higher in recent injuries than chronic ones. According to Saal and colleagues, PLA2 is not part of an autoimmune response. It is endogenous to the disc and released by herniation.

Here's another example that supports the endogenous explanation. As we have seen, when the nucleus of a disc is applied to nerve root tissue, the axons and myelin sheaths degenerate and vascular permeability increases. Olmarker and colleagues deduced that a likely cause of such a pattern of damage was the inflammatory cytokine TNF (261). They proceeded to identify TNF in nucleus material and, confirming their theory, found that TNF inhibitors protected nerves against damage from nucleus material (66,67,262). So, along with PLA2, TNF seems to be an irritant to nerve roots that is endogenous to the disc. An interesting recent case study used a sophisticated imaging technique to detect TNF in the disc herniation of a patient with radicular pain. After a successful operation, the TNF was gone (263). (Unfortunately, anti-TNF drugs don't seem to work clinically, in people with 'sciatica' (264). It seems that chemical irritation is too complex a process to be thwarted by removing just one factor, or indeed this mechanism may not be predominant in all patients.)

One last example of the endogenous explanation. Discs also contain a protein called monocyte chemoattractant protein. Monocyte chemoattractant protein is inactive until it is exposed to the body outside the disc whereupon it does what its name implies and attracts immune cells called monocyctes (265). Those monocytes then differentiate into another type of immune cells called macrophages. Macrophages break down and digest surrounding tissues and secrete inflammatory mediators and, to create a positive feedback loop, also produce more monocyte chemoattractant protein themselves.

So you can see that according to the endogenous explanation, there are a great many substances inside a degenerating disc that can cause chemical irritation to the nerve root. And they build up slowly over time, to be released onto nerve roots if a disc herniates.

The autoimmune explanation

The interior of the human intervertebral disc is cut off from the outside world of the rest of the body (266,267). Except for the outer few millimetres, the disc has no nerve supply and no blood supply. Nerves and blood vessels cannot infiltrate through the tightly bound, high-pressure fortress of the annulus. And, for added protection, this fortress is patrolled by cell-destroying molecules.

This makes the interior of a healthy disc one of a select few sites in the body, along with the insides of eyeballs, the insides of testes and placenta, that have what is called immune privilege.

If the outer annulus cracks or the disc herniates, the fortress maintaining this immune privilege is breached (268). This is the first time in a person's lifetime that their disc material is presented to their immune cells. Those immune cells (T-cells, which we saw before when we looked at neuroinflammation) are unable to differentiate this new substance from a pathogen or toxin. So they initiate a protective inflammatory reaction, recruiting other immune cells to the area.

The autoimmune explanation was first proposed in the 1960s. It was supported early on by an experiment in which nucleus material from the discs of rabbits was applied to the animals' ears, which caused inflammation and increased vascularity which, the researchers noticed, looked like an autoimmune reaction (269). This likely autoimmune response peaked at four days and lasted about three weeks. Other similar animal experiments followed. In 1977, Marshall and colleagues, for example, harvested nucleus material from a human cadaver and injected it to the preserved lung of a guinea pig (270). They observed a 'severe reaction' of bronchoconstriction and edema in the lung. They then tried the same thing on preserved guinea pig intestines, which reacted by constricting repeatedly. Finally, to confirm that they were observing an autoimmune reaction, the authors measured serum levels of circulating antibodies in their patients. Their results were conflicting, but Marshall and colleagues took them as a suggestion that an autoimmune reaction was present.

The autoimmune explanation partly accounts for why radicular pain can go on for so much longer than normal tissue healing times. And, given that autoimmune reactions recruit lots of macrophages (eating-up cells), it partly explains why discs usually reduce in size over time. Extruded herniations that break out past the posterior longitudinal ligament contain more markers of an autoimmune reaction and also reduce in size more, so that fits the picture (216,265,266).

But some people think the autoimmune reaction is not that important. Jonsson and Olmarker point out that an autoimmune reaction would take time to build up, but disc material sometimes seems to cause problems to neural tissue very quickly (5). And, they say, while it's true that a person's immune system might not have seen their nucleus before, it has of course seen that person's genetic signature before. According to Jonsson and Olmarker, that should prevent it from starting an autoimmune reaction.

Key points on how disc material chemically irritates a nerve root complex:

- The endogenous explanation states that disc material, especially from ageing or degenerating discs, can be intrinsically irritating if in close proximity to nerve roots.

- The autoimmune explanation states that internal disc material is seen as foreign by the immune system, which triggers an immune-inflammatory reaction that can irritate nerve roots.
- Both explanations are probably true, although there is a bit more controversy about the autoimmune explanation.

25

DISC RESORPTION

Let's finish by looking at what happens to a herniation in the long term.

Most herniations change size, and most get smaller. How does this happen? For a while, people thought maybe the herniation gets sucked back into the disc, or that maybe it dries out and disintegrates (217). But recent evidence shows that it's all thanks to the immune system (265). As we saw earlier, the immune system drapes disc herniations with soft, bloody granulation tissue. Inside that granulation tissue are macrophages (271). Macrophages gobble up herniated disc material so that the herniation slowly gets smaller over time (272,273). This is called resorption.

A big disc herniation is more likely to resorb than a small disc herniation because it has more surface area for the immune system to work on. And if the disc herniation has broken past the posterior longitudinal ligament then it is even more likely to resorb, because in the spinal canal it is exposed to more blood vessels and thus more macrophages (a nice example of why the immune-inflammatory response can be seen as a good thing, despite the pain) (216,254). Illustrating that bigger is better when it comes to disc resorption, 96%

of sequestrations and 70% of extrusions are at least partly resorbed, compared to 41% of protrusions (217).

This phenomenon is why, even when surgeons are concerned about cauda equina syndrome, some choose not to operate on people with massive central herniations in the absence of clinical symptoms or signs. In one study, fifteen such people were observed for two years; everyone's herniation substantially resorbed and none developed cauda equina syndrome (274).

As well as its size, the composition of a disc herniation also has some bearing on whether or not it is resorbed. More macrophages are found in nucleus material than annulus (275). This might be because nucleus material swells dramatically after herniation, which presents a greater target for granulation tissue. As you would expect, cartilage and bone seem to be more resistant to resorption (226).

How long does resorption take? Well, macrophages are quite slow starters. Studies that look for resorption within the first six weeks or so do not find much (276). After that, macrophages pick up the pace. They seem to mostly finish their work within the year.

That said, not all herniations resorb. Studies that look over a period of many years find that some just stick around (276). Some long-termers are probably dormant, no longer much of a chemical irritant, and some even calcify. So, if you see a disc herniation on MRI, it may well resorb but it may well not.

In fact, a disc herniation might get even bigger over time! This might be a bit surprising if you tend to imagine that herniations spurt out all at once. But some are in fact gradual (277). It's hard to say how many, but it seems like about one fifth to one third of relatively new herniations will actually get bigger over the weeks and months to come (216,251). Smaller herniations are more likely to get bigger than already-big ones. This includes some contained herniations that subsequently break through the posterior longitudinal ligament to become uncontained (216).

Key points on disc resorbtion:

- Many disc herniations get smaller over time, as they are resorbed by macrophages.
- That said, some herniations are gradual, getting bigger for a time before they start to resorb.
- Extrusions are more likely to resorb than protrusions because they are more exposed to the immune system.

26

SIDE NOTE: FORAMINAL STENOSIS

In 1927, the Italian physician Vittorio Putti delivered a lecture at the University of Liverpool called 'New Conceptions in the Pathogenesis of Sciatic Pain' (10). In the lecture, Putti gave one of the earliest descriptions of foraminal stenosis:

> '*The diseased [intervertebral] joint, by its swelling and deformity, changes the shape and capacity of the foramen, thus irritating and compressing the nerve within it. [...] The intervertebral foramen constitutes a critical region, or as Sicard has happily named it, 'carrefour de la douleur' -i.e., the crossroads of neuralgia. Any condition which modifies in the slightest degree the contents, or the container, at once induces a painful reaction, which is referred distally to the sciatic nerve.*'

Putti was speaking less than a decade before Mixter and Barr would publish their paper linking sciatica with the disc herniation (182). That momentous paper ushered in the 'dynasty of the disc' which overshadowed Putti's observation (183). Foraminal stenosis slipped from public consciousness. Writing in 1971, one author called the foramen 'the hidden zone' as it could not be seen easily using imaging techniques used at the time, or by the surgeon intra-operatively (278). One large study in 1981 found that 60% of patients who had not benefited from decompression surgery for radicular pain in fact had a foraminal stenosis that had not been identified (279).

As we have seen, the nerve root complex occupies the upper third of the space of the intervertebral foramen, hugging the ceiling of the passageway. And much of the rest of the space in the foramen is taken up by ligaments, blood vessels and fatty tissue (280).

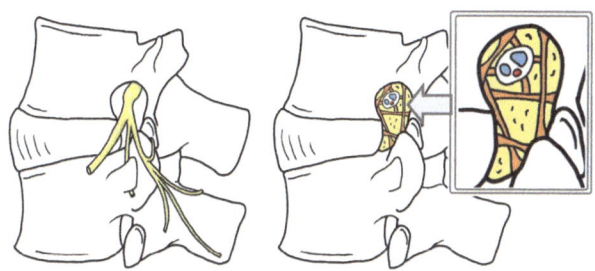

Side view of the intervertebral foramen. The nerve root is in cross section (note the dorsal and slightly smaller ventral root, accompanied by the distal radicular artery). The rest of the foramen contains a latticework of intra-foraminal ligaments and cushioning fatty tissue. From this angle, it is easy to see how bony overgrowth, ligament hypertrophy, neural edema, disc bulging and narrowing etc. can all cause stenosis (i.e. narrowing) of the foramen.

What causes foraminal stenosis?

What causes the foramen to narrow and encroach on its contents? Its floor and ceiling narrow when the disc loses its height through age or herniation, causing the vertebrae above and below to close in on one another (281). One cadaveric study found that once a disc was 4mm or less in height, it began to significantly compress the nerve root (282). The ceiling of the foramen can further impinge on the root if it grows bone spurs, or osteophytes. Osteophytes can be thought of like calluses which actually increase the resilience of the intervertebral joint with age (283). That they increase foraminal stenosis is an unwanted consequence for some people. And as for the walls of the foramen, they narrow when, with age and disc degeneration, the superior facet subluxes and the ligamentum flavum buckles.

About 75% of the time, foraminal stenosis is located around the L5 nerve root (284). This is because degenerative changes are more common at the L5/S1 intervertebral joint and because the exiting dorsal root ganglion is particularly large at this level.

Foraminal stenosis is a good example of how mechanical compression can be the sole primary cause of radicular pain, without any primary chemical injury. It is likely that the 'mini compartment syndrome' that we saw earlier has a significant role here. By extension, and given that the foramen is a passageway for blood vessels as well as the nerve root complex, a lack of oxygen is likely to contribute to much of the early pain. Because the insult to the nerve root complex is more later-ally placed than most disc lesions, the dorsal root ganglion is likely to be directly involved, which might account for why foraminal stenosis can be so painful. On the other hand, foraminal stenosis does occur slowly, and as we have seen nerves are more resilient to slowly-applied pressure. So, asymptomatic foraminal stenosis is common. Even the dorsal root ganglion is frequently found to have been squashed on autopsies of people who never had radicular pain (39).

No one knows exactly how common foraminal stenosis is. It seems to account for a small but significant minority of radicular pain cases. In the recent ATLAS study, of all people with sciatica, about half had nerve root compression, and of those, 18% had stenosis (135). But the authors did not specify whether that was of the canal or foramen. One cadaveric study found that of 100 foramina, 21 were stenotic enough to compress the nerve root complex within (282). But we don't know how many of these cases caused symptoms. In his 1984 paper, the physician Richard Porter reported that 11% of the patients attending his back pain clinic had foraminal stenosis (219).

What does it look like, clinically?

Porter was a masterful observer of clinical patterns. He described four key features of foraminal stenosis (219).

- Firstly, it causes severe, constant radicular pain.
- Secondly, the pain is unrelieved by bed rest. Porter wrote that his patients 'would pace the floor at night and seek to keep changing position when in a chair'. (These first two features were confirmed recently by Yamada and colleagues (285)).
- Thirdly, the straight leg raise test is often negative. 60% of Porter's patients with foraminal stenosis had a straight leg raise to ninety degrees or more. They could bend and touch their shins with little difficulty. On the other hand, back extension was usually limited, because it further narrows the space of the foramen. A quarter of Porter's patients could not extend their backs at all.
- Finally, patients with lumbar foraminal stenosis are usually older than patients with predominant disc-related sciatica. Indeed, foraminal stenosis can be found twice as often in patients with sciatica aged 65-85 than in those aged 35-55 (286).

Additionally, although he did not include it in his four criteria, Porter and other subsequent authors have all observed that back pain almost always accompanies foraminal stenosis. In fact, patients often described that back pain started first.

27

SUMMARY OF PART III

As we did after the first half of this book, let's summarise what we've just seen.

Where are nerve roots injured?

The nerve roots in the lower back leave the dural sac and swoop downwards and outwards until they leave the spine through the intervertebral foramen. They are most commonly injured at the most proximal part of their journey, just after they've left the dural sac and while they're still 'transiting' downward. This is the only point at which the roots, in particular the L5 and S1 root, pass directly over an intervertebral disc. And the part of the disc they pass over, the paracentral part, also happens to be a frequent site for herniations.

Roots are held relatively in place, on the disc-side of the spinal canal, by the dural sac at their proximal end a number of ligaments at their distal end. This means that even small herniations can contact a nerve root and the nerve root has little ability to 'escape'.

What does a healthy disc look like?

Healthy discs are composed of tough but porous endplates, which sit on the top and bottom of the disc; a stiff, multi-layered, tire-like annulus, which circles the edges; and a soft and watery nucleus, which is cloistered in the middle.

The disc is best thought of as a force-distributor. The watery nucleus takes downward force and distributes it outward, into the rings of the annulus. When we bend, this downward force is pushed backwards, which puts a lot of pressure on the back part of the annulus.

How do disc herniations happen?

Although the disc can adapt and get stronger like any other tissue, if the backward force of bending is too great or too regular, then the annulus can fail and develop a fissure. This fissure is one way that herniations begin. The nucleus progressively 'noses' its way through the annulus until it breaks through the final layer, herniating.

That said, the annulus is very strong, so often it will not break but instead tear off a piece of endplate, and even bone. Either way, nucleus material still herniates from the disc.

Of course, disc herniations are about more than bending repeatedly. Genetic inheritance makes some discs more susceptible than others, as does general lifestyle health.

Why doesn't the disc 'heal up' against these fissures?

The disc's microenvironment is delicately balanced. The main reason for this is that, because it's such a high pressure environment, the inner part of the disc doesn't have blood vessels, which means it cannot easily get oxygen and nutrients in or waste products out. As a consequence, discs 'age' earlier than many other parts of the body, where 'ageing' means that its tissues begin to degrade faster than they can renew themselves.

All this means that when a disc sustains a fissure, it cannot heal as readily as other tissues. In fact, in its attempts to respond, the disc can fall into a highly disordered metabolic pattern called disc degeneration.

What are the different types of herniation?

A disc bulge is a generalised displacement of much of the circumference of the disc.

A disc herniation is a localised displacement.

If the herniation is still contained within the annulus, it's a protrusion. If it's broken through, it's an extrusion. Sometimes, it breaks through the posterior longitudinal ligament too, in which case it has become uncontained. Finally, a free fragment of disc herniation is called a sequestration.

You can also define a herniation by its location. Looking down on the disc, the most common location is the 5 or 7 o'clock herniation, called para-central.

Why is disc material a chemical irritant to nerve roots?

The 'endogenous explanation' says that discs that herniate are often also degenerating, and so contain lots of pro-inflammatory chemicals. By contrast, the 'autoimmune explanation' says that it's because the nucleus is normally sealed off from the immune system, and is therefore identified as a foreign body once it herniates.

It's likely to be a bit of both.

Do disc herniations resorb?

Yes, in particular extrusions because they're more exposed to the immune system. That said, some disc herniations are progressive, so it's not altogether true to say that every herniation will just get smaller.

In any case, resorption is only slightly relevant, clinically. After all, the size of a herniation is rendered less important than one might think by the fact that roots are held relatively in place, close to the disc. On top of that, there are so many other factors that determine the clinical picture, not least the amount of chemical irritation provided by the disc and the immune system's response to it.

All told, disc herniations don't cause sciatica the way a poke in the eye causes eye pain! Instead, disc herniations are more analogous to putting some chilli powder on your finger and *then* poking (or just touching) your eye. Structural and mechanical factors are accompanied by a chemical irritant, which stimulates an active inflammatory process.

And finally, not all radicular pain is associated with disc herniations. For instance, foraminal stenosis can also irritate the nerve roots.

BIBLIOGRAPHY

1. Milette P. Radiculopathy, radicular pain, radiating pain, referred pain: what are we really talking about? Radiology. 1994;192:280-2.
2. International Association for the Study of Pain. Pain terms a current list with definitions and notes on usage. Pain. 1986;24:215-21.
3. Kellgren J. On the distribution of pain arising from deep somatic structures with charts of segmental pain areas. Clin Sci. 1939;4(1):35-46.
4. Bogduk N. On the definitions and physiology of back pain, referred pain, and radicular pain. Pain. 2009 Dec;147(1):17-9.
5. Taylor CS, Coxon AJ, Watson PC, Greenough CG. Do L5 and S1 Nerve Root Compressions Produce Radicular Pain in a Dermatomal Pattern?. Spine. 2013 May;38(12):995-8.
6. Albert HB, Hansen JK, Søgaard H, Kent P. Where do patients with MRI-confirmed single-level radiculopathy experience pain, and what is the clinical interpretability of these pain patterns? A cross-sectional diagnostic accuracy study. Chiropr Man Ther. 2019 Dec;27(1).
7. Furman MB, Johnson SC. Induced lumbosacral radicular symptom referral patterns: a descriptive study. Spine J. 2019 Jan;19(1):163-70.
8. Murphy DR, Hurwitz EL, Gerrard JK, Clary R. Pain patterns and descriptions in patients with radicular pain: Does the pain necessarily follow a specific dermatome? Chiropr Osteopat. 2009 Dec;17(1).
9. Gifford L. Acute low cervical nerve root conditions: symptom presentations and pathobiological reasoning. Man Ther. 2001 May;6(2):106-15.
10. Putti V. New conceptions in the pathogenesis of sciatic pain. Lancet. 1927;2:54-60.
11. Fairbank JCT. An archaic term. BMJ. 2007 Jul;335(7611):112.
12. Lin CWC, Verwoerd AJH, Maher CG, Verhagen AP, Pinto RZ, Luijsterburg P A J, Hancock MJ. How is radiating leg pain defined in randomized controlled trials of conservative treatments in primary care? A systematic review. Eur J Pain. 2014;18(4):455-64.
13. International Association for the Study of Pain. Classification of Chronic Pain, Second Edition (Revised) [Internet]. International Association for the Study of Pain (IASP). [cited 2022 Sep 2]. Available from: https://www.iasp-pain.org/publications/free-ebooks/classification-of-chronic-pain-second-edition-revised/
14. Cohen M, Wall E, Kerber C, Abitbol J, Garfin. The anatomy of the cauda equina on CT scans and MRI. J Bone Joint Surg Br. 1991 May;73-B(3):381-4.
15. Hu P, Bembrick AL, Keay KA, McLachlan EM. Immune cell involvement in dorsal root ganglia and spinal cord after chronic constriction or transection of the rat sciatic nerve. Brain Behav Immun. 2007 Jul;21(5):599-616.

16. Butler DS. The sensitive nervous system. Adelaide: Noigroup Publications; 2009.

17. Rauschning W. Pathoanatomy of lumbar disc degeneration and stenosis. Acta Orthop Scand. 1993 Jan;64(251):3–12.

18. Sunderland S. Meningeal-neural relations in the intervertebral foramen. J Neurosurg. 1974 Jun;40(6):756–63.

19. Söderberg L. Prognosis in Conservatively Treated Sciatica. Acta Orthop Scand. 1956 Dec;27(21):3–127.

20. Smyth M, Wright V. Sciatica and the Intervertebral Disc: An Experimental Study. J Bone Jt Surg. 1958 Dec;40(6):1401–18.

21. Shi J gang, Xu X ming, Sun J chuan, Wang Y, Kong Q jie, Shi G dong. Theory of Bowstring Disease: Diagnosis and Treatment Bowstring Disease: Theory of Bowstring Disease. Orthop Surg. 2019 Feb;11(1):3–9.

22. Falconer MA, McGeorge M, Begg AC. Observations on the cause and mechanism of symptom-production in sciatica and low-back pain. J Neurol Neurosurg Psychiatry. 1948 Feb;11(1):13–26.

23. Sarazin L, Chevrot A, Pessis E, Minoui A, Drape JL, Chemla N, et al. Lumbar Facet Joint Arthrography with the Posterior Approach. RadioGraphics. 1999 Jan;19(1):93–104.

24. Schmidt BL, Hamamoto DT, Simone DA, Wilcox GL. Mechanism of Cancer Pain. Mol Interv. 2010 Jun;10(3):164–78.

25. Lindblom K, Rexed B. Spinal Nerve Injury in Dorso-Lateral Protrusions of Lumbar Disks. J Neurosurg. 1948 Sep;5(5):413–32.

26. Lundborg G, Myers R, Powell H. Nerve compression injury and increased endoneurial fluid pressure: a 'miniature compartment syndrome'. J Neurol Neurosurg Psychiatry. 1983 Dec;46(12):1119–24.

27. Lundborg G, Gelberman RH, Minteer-Convery M, Lee YF, Hargens AR. Median nerve compression in the carpal tunnel—Functional response to experimentally induced controlled pressure. J Hand Surg. 1982 May;7(3):252–9.

28. Olmarker K, Rydevik B, Holm S, Bagge U. Effects of experimental graded compression on blood flow in spinal nerve roots. A vital microscopic study on the porcine cauda equina. J Orthop Res. 1989 Nov;7(6):817–23.

29. Takahashi K, Shima I, Porter RW. Nerve Root Pressure in Lumbar Disc Herniation: Spine. 1999 Oct;24(19):2003.

30. Sunderland S. The nerve lesion in the carpal tunnel syndrome. J Neurol Neurosurg Psychiatry. 1976 Jul 1;39(7):615–26.

31. Yoshizawa H, Kobayashi S, Morita T. Chronic Nerve Root Compression: Pathophysiologic Mechanism of Nerve Root Dysfunction. Spine. 1995 Feb;20(4):397–407.

32. Rydevik B, Brown MD, Lundborg G. Pathoanatomy and pathophysiology of nerve root compression. Spine. 1984 Feb;9(1):7–15.

33. Grieve GP. Common vertebral joint problems. Edinburgh: Churchill Livingstone; 1981.

34. Takata K, Inoue S, Takahashi K, Ohtsuka Y. Swelling of the cauda equina in patients

who have herniation of a lumbar disc. A possible pathogenesis of sciatica. J Bone Joint Surg Am. 1988 Mar;70(3):361–8.

35. Toyone T, Takahashi K, Kitahara H, Yamagata M, Murakami M, Moriya H. Visualisation of symptomatic nerve roots. Prospective study of contrast-enhanced MRI in patients with lumbar disc herniation. J Bone Joint Surg Br. 1993 Jul;75-B(4):529–33.

36. Kobayashi S, Mwaka ES, Meir A, Uchida K, Kokubo Y, Takeno K, et al. Changes in Blood Flow, Oxygen Tension, Action Potentials, and Vascular Permeability Induced by Arterial Ischemia or Venous Congestion on the Lumbar Dorsal Root Ganglia in Dogs. J Neurotrauma. 2009 Jul;26(7):1167–75.

37. Pronin S, Koh CH, Bulovaite E, Macleod MR, Statham PF. Compressive Pressure Versus Time in Cauda Equina Syndrome. Spine. 2019 Aug 1;44(17):1238–47.

38. Buttermann GR. Treatment of Lumbar Disc Herniation: Epidural Steroid Injection Compared with Discectomy. J Bone Joint Surg Am. 2004 Apr;86(4):670-9

39. Rauschning W. Normal and pathologic anatomy of the lumbar root canals. Spine. 1987 Dec;12(10):1008–19.

40. Akuthota V, Marshall B, Boimbo S, Osborne MC, Garvan CS, Garvan GJ, et al. Clinical Course of Motor Deficits from Lumbosacral Radiculopathy Due to Disk Herniation. PM&R. 2019;11(8):807–14.

41. Overdevest GM, Vleggeert-Lankamp CLAM, Jacobs WCH, Brand R, Koes BW, Peul WC. Recovery of motor deficit accompanying sciatica—subgroup analysis of a randomized controlled trial. Spine J. 2014 Sep;14(9):1817–24.

42. Balaji VR, Chin KF, Tucker S, Wilson LF, Casey AT. Recovery of severe motor deficit secondary to herniated lumbar disc prolapse: is surgical intervention important? A systematic review. Eur Spine J. 2014 Sep;23(9):1968–77.

43. Kobayashi S, Takeno K, Yayama T, Awara K, Miyazaki T, Guerrero A, et al. Pathomechanisms of sciatica in lumbar disc herniation: effect of periradicular adhesive tissue on electrophysiological values by an intraoperative straight leg raising test. Spine. 2010 Oct 15;35(22):2004–14.

44. Kobayashi S, Yoshizawa H, Yamada S. Pathology of lumbar nerve root compression Part 1: Intraradicular inflammatory changes induced by mechanical compression. J Orthop Res. 2004 Jan;22(1):170–9.

45. Suri P, Rainville J, Gellhorn A. Predictors of Patient-Reported Recovery From Motor or Sensory Deficits Two Years After Acute Symptomatic Lumbar Disk Herniation. PM&R. 2012 Dec;4(12):936-944.

46. Ramer MS, McMahon SB, Priestley JV. Axon regeneration across the dorsal root entry zone. Prog Brain Res. 2001;132:621–39.

47. O'Brien JP, Mackinnon SE, MacLean AR, Hudson AR, Dellon AL, Hunter DA. A model of chronic nerve compression in the rat. Ann Plast Surg. 1987 Nov;19(5):430–5.

48. Gupta R, Rummler L, Steward O. Understanding the Biology of Compressive Neuropathies: Clin Orthop. 2005 Jul;NA;436:251–60.

49. Jancalek R, Dubovy P. An experimental animal model of spinal root compression

syndrome: an analysis of morphological changes of myelinated axons during compression radiculopathy and after decompression. Exp Brain Res. 2007 Apr 23;179(1):111–9.

50. Matsui T, Takahashi K, Moriya M, Tanaka S, Kawahara N, Tomita K. Quantitative Analysis of Edema in the Dorsal Nerve Roots Induced by Acute Mechanical Compression: Spine. 1998 Sep;23(18):1931–6.

51. Pedowitz RA, Garfin SR, Massie JB, Hargens AR, Swenson MR, Myers RR, et al. Effects of magnitude and duration of compression on spinal nerve root conduction. Spine. 1992 Feb;17(2):194–9.

52. Schoenfeld AJ, Bono CM. Does Surgical Timing Influence Functional Recovery After Lumbar Discectomy? A Systematic Review. Clin Orthop. 2015 Jun;473(6):1963–70.

53. Olmarker K, Rydevik B, Holm S. Edema formation in spinal nerve roots induced by experimental, graded compression. An experimental study on the pig cauda equina with special reference to differences in effects between rapid and slow onset of compression. Spine. 1989 Jun;14(6):569–73.

54. Kikuchi S, Konno S, Kayama S, Sato K, Olmarker K. Increased resistance to acute compression injury in chronically compressed spinal nerve roots. An experimental study. Spine. 1996 Nov 1;21(22):2544–50.

55. Cooper RG, Freemont AJ, Hoyland JA, Jenkins JP, West CG, Illingworth KJ, et al. Herniated intervertebral disc-associated periradicular fibrosis and vascular abnormalities occur without inflammatory cell infiltration. Spine. 1995 Mar 1;20(5):591–8.

56. Hoyland J, Freemont A, Jayson M. Intervertebral foramen venous obstruction. A cause of periradicular fibrosis? Spine. 1989;14(6):558-68

57. Shinotsuka N, Denk F. Fibroblasts: the neglected cell type in peripheral sensitisation and chronic pain? A review based on a systematic search of the literature. BMJ Open Sci. 2022 Jan 18;6(1):e100235.

58. Kuslich S, Ulstrom S, Michael S. The tissue origin of low back pain and sciatica: a report of pain response to tissue stimulation during operations on the lumbar spine using local anesthesia. Orthop Clin North Am. 1991 Apr 1;22(2):181–7.

59. Helm S, Racz GB, Gerdesmeyer L, Justiz R, Hayek SM, Kaplan ED, et al. Percutaneous and Endoscopic Adhesiolysis in Managing Low Back and Lower Extremity Pain: A Systematic Review and Meta-analysis. Pain Physician. 2016;19;E245-E281

60. el Barzouhi A, Vleggeert-Lankamp CLAM, Lycklama à Nijeholt GJ, Van der Kallen BF, van den Hout WB, Jacobs WCH, et al. Magnetic Resonance Imaging in Follow-up Assessment of Sciatica. N Engl J Med. 2013 Mar 14;368(11):999–1007.

61. Olmarker K, Nordborg C, Larsson K, Rydevik B. Ultrastructural Changes in Spinal Nerve Roots Induced by Autologous Nucleus Pulposus. Spine. 1996 Feb 15;21(4):411–4.

62. Watkins LR, Maier SF. Beyond Neurons: Evidence That Immune and Glial Cells Contribute to Pathological Pain States. Physiol Rev. 2002 Jan 10;82(4):981–1011.

63. Albrecht DS, Granziera C, Hooker JM, Loggia ML. In Vivo Imaging of Human Neuroinflammation. ACS Chem Neurosci. 2016 Apr 20;7(4):470–83.

64. Colloca L, Ludman T, Bouhassira D, Baron R, Dickenson AH, Yarnitsky D, et al. Neuropathic pain. Nat Rev Dis Primer. 2017 Feb 16;3:17002.
65. Ellis A, Bennett DLH. Neuroinflammation and the generation of neuropathic pain. Br J Anaesth. 2013 Jul;111(1):26–37.
66. Olmarker K, Byröd G, Cornefjord M, Nordborg C, Rydevik B. Effects of methyl-prednisolone on nucleus pulposus-induced nerve root injury. Spine. 1994 Aug 15;19(16):1803–8.
67. Yabuki S, Onda A, Kikuchi S, Myers RR. Prevention of compartment syndrome in dorsal root ganglia caused by exposure to nucleus pulposus. Spine. 2001 Apr 15;26(8):870–5.
68. Yabuki S, Kikuchi S, Olmarker K, Myers RR. Acute effects of nucleus pulposus on blood flow and endoneurial fluid pressure in rat dorsal root ganglia. Spine. 1998 Dec 1;23(23):2517–23.
69. Murata Y, Rydevik B, Takahashi K, Larsson K, Olmarker K. Incision of the Intervertebral Disc Induces Disintegration and Increases Permeability of the Dorsal Root Ganglion Capsule: Spine. 2005 Aug;30(15):1712–6.
70. Murata Y, Nannmark U, Rydevik B, Takahashi K, Olmarker K. Nucleus Pulposus-Induced Apoptosis in Dorsal Root Ganglion Following Experimental Disc Herniation in Rats: Spine. 2006 Feb;31(4):382–90.
71. Parisien M, Lima LV, Dagostino C, El-Hachem N, Drury GL, Grant AV, et al. Acute inflammatory response via neutrophil activation protects against the development of chronic pain. Sci Transl Med. 2022 May 11;14(644):eabj9954.
72. Antonelli M, Kushner I. It's time to redefine inflammation. FASEB J. 2017 May;31(5):1787–91.
73. Andrasinova T, Kalikova E, Kopacik R, Srotova I, Vlckova E, Dusek L, et al. Evaluation of the Neuropathic Component of Chronic Low Back Pain: Clin J Pain. 2019 Jan;35(1):7–17.
74. Drummond PD, Morellini N, Visser E, Finch PM. Parallels between lumbosacral radiculopathy and complex regional pain syndrome: α1-adrenoceptor upregulation, reduced dermal nerve fibre density, and hemisensory disturbances in postsurgical sciatica. PAIN. 2019 Aug;160(8):1891–900.
75. Devor M. Unexplained peculiarities of the dorsal root ganglion: Pain. 1999 Aug;82:S27–35.
76. Rasmussen VF, Karlsson P, Drummond PD, Schaldemose EL, Terkelsen AJ, Jensen TS, et al. Bilaterally Reduced Intraepidermal Nerve Fiber Density in Unilateral CRPS-I. Pain Med Malden Mass. 2018 Oct 1;19(10):2021–30.
77. Farrell SF, Sterling M, Irving-Rodgers H, Schmid AB. Small fibre pathology in chronic whiplash-associated disorder: A cross-sectional study. Eur J Pain Lond Engl. 2020 Jul;24(6):1045–57.
78. London D, Birkenfeld B, Thomas J, Avshalumov M, Mogilner AY, Falowski S, et al. A broad and variable lumbosacral myotome map uncovered by foraminal nerve root stimulation. J Neurosurg Spine. 2022;37(5):680-686.
79. McMahon SB, editor. Wall and Melzack's textbook of pain. 6th ed. Philadelphia, PA:

Elsevier/Saunders; 2013.
80. Teixeira MJ, Almeida DB, Yeng LT. Concept of acute neuropathic pain. The role of nervi nervorum in the distinction between acute nociceptive and neuropathic pain. Rev Dor. 2016;17(1):5-10.
81. Hida S, Naito M, Kubo M. Intraoperative Measurements of Nerve Root Blood Flow During Discectomy for Lumbar Disc Herniation: Spine. 2003 Jan;28(1):85–90.
82. Tampin B, Slater H, Jacques A, Lind CRP. Association of quantitative sensory testing parameters with clinical outcome in patients with lumbar radiculopathy undergoing microdiscectomy. Eur J Pain Lond Engl. 2020 Aug;24(7):1377–92.
83. Mansilha A, Sousa J. Pathophysiological Mechanisms of Chronic Venous Disease and Implications for Venoactive Drug Therapy. Int J Mol Sci. 2018 Jun 5;19(6):1669.
84. Devor M. Ectopic Generators. In: The Senses: A Comprehensive Reference. Elsevier; 2008. p. 83–8.
85. Finnerup NB, Kuner R, Jensen TS. Neuropathic pain: From mechanisms to treatment. Physiol Rev. 2021;101(1):259-301
86. Mahn F, Hüllemann P, Gockel U, Brosz M, Freynhagen R, Tölle TR, et al. Sensory Symptom Profiles and Co-Morbidities in Painful Radiculopathy. Priller J, editor. PLoS ONE. 2011 May 9;6(5):e18018.
87. Goodwin G, Bove GM, Dayment B, Dilley A. Characterizing the Mechanical Properties of Ectopic Axonal Receptive Fields in Inflamed Nerves and Following Axonal Transport Disruption. Neuroscience. 2020 Mar 1;429:10–22.
88. Howe JF, Loeser JD, Calvin WH. Mechanosensitivity of dorsal root ganglia and chronically injured axons: A physiological basis for the radicular pain of nerve root compression: Pain. 1977 Feb;3(1):25–41.
89. Song XJ, Hu SJ, Greenquist KW, Zhang JM, LaMotte RH. Mechanical and Thermal Hyperalgesia and Ectopic Neuronal Discharge After Chronic Compression of Dorsal Root Ganglia. J Neurophysiol. 1999 Dec 1;82(6):3347–58.
90. Hogan Q. Labat Lecture: The Primary Sensory Neuron: Where it is, What it Does, and Why it Matters. Reg Anesth Pain Med. 2010;35(3):306–11.
91. Lindahl O. Hyperalgesia of the Lumbar Nerve Roots in Sciatica. Acta Orthop Scand. 1966 Jan;37(4):367–74.
92. Jensen OK, Nielsen CV, Sørensen JS, Stengaard-Pedersen K. Back pain was less explained than leg pain: a cross-sectional study using magnetic resonance imaging in low back pain patients with and without radiculopathy. BMC Musculoskelet Disord. 2015 Dec;16:374.
93. Zhang JM, An J. Cytokines, Inflammation, and Pain: Int Anesthesiol Clin. 2007;45(2):27–37.
94. Ji RR, Chamessian A, Zhang YQ. Pain Regulation by Non-neuronal Cells and Inflammation. Science. 2016 Nov 4;354(6312):572–7.
95. Chen O, Donnelly CR, Ji RR. Regulation of pain by neuro-immune interactions between macrophages and nociceptor sensory neurons. Curr Opin Neurobiol. 2020 Jun;62:17–25.
96. Shubayev VI, Kato K, Myers RR. Cytokines in Pain. In: Kruger L, Light AR, editors.

Translational Pain Research: From Mouse to Man. Boca Raton (FL): CRC Press/Taylor & Francis; 2010

97. Moalem G, Tracey DJ. Immune and inflammatory mechanisms in neuropathic pain. Brain Res Rev. 2006 Aug;51(2):240–64.

98. Yan J, Zou K, Liu X, Hu S, Wang Q, Miao X, et al. Hyperexcitability and sensitization of sodium channels of dorsal root ganglion neurons in a rat model of lumber disc herniation. Eur Spine J Off Publ Eur Spine Soc Eur Spinal Deform Soc Eur Sect Cerv Spine Res Soc. 2016 Jan;25(1):177–85.

99. Dilley A, Bove GM. Disruption of axoplasmic transport induces mechanical sensitivity in intact rat C-fibre nociceptor axons: Axoplasmic transport block causes axonal mechanical sensitivity. J Physiol. 2008 Jan 15;586(2):593–604.

100. Satkeviciute I, Goodwin G, Bove GM, Dilley A. Time course of ongoing activity during neuritis and following axonal transport disruption. J Neurophysiol. 2018 May;119(5):1993–2000.

101. Bove GM, Dilley A. A lesson from classic British literature. The Lancet. 2019 Mar;393(10178):1297–8.

102. Otoshi K, Kikuchi S, Konno S, Sekiguchi M. The Reactions of Glial Cells and Endoneurial Macrophages in the Dorsal Root Ganglion and Their Contribution to Pain-Related Behavior After Application of Nucleus Pulposus Onto the Nerve Root in Rats. Spine. 2010 Feb; 35(3):264–71.

103. Takebayashi T, Cavanaugh JM, Cüneyt Özaktay A, Kallakuri S, Chen C. Effect of Nucleus Pulposus on the Neural Activity of Dorsal Root Ganglion: Spine. 2001 Apr;26(8):940–4.

104. Brisby H, Hammar I. Thalamic activation in a disc herniation model. Spine. 2007 Dec 1;32(25):2846–52.

105. Nilsson E, Brisby H, Rask K, Hammar I. Mechanical Compression and Nucleus Pulposus Application on Dorsal Root Ganglia Differentially Modify Evoked Neuronal Activity in the Thalamus. BioResearch Open Access. 2013 Jun;2(3):192–8.

106. Ohtori S, Inoue G, Koshi T, Ito T, Doya H, Saito T, et al. Up-regulation of acid-sensing ion channel 3 in dorsal root ganglion neurons following application of nucleus pulposus on nerve root in rats. Spine. 2006 Aug 15;31(18):2048–52.

107. Amaya F, Samad TA, Barrett L, Broom DC, Woolf CJ. Periganglionic inflammation elicits a distally radiating pain hypersensitivity by promoting COX-2 induction in the dorsal root ganglion: Pain. 2009 Mar;142(1):59–67.

108. Cavaletti G, Alberti P, Argyriou AA, Lustberg M, Staff NP, Tamburin S. Chemotherapy-induced peripheral neurotoxicity: A multifaceted, still unsolved issue. J Peripher Nerv Syst. 2019;24(S2):S6–12.

109. McLachlan EM, Jänig W, Devor M, Michaelis M. Peripheral nerve injury triggers noradrenergic sprouting within dorsal root ganglia. Nature. 1993 Jun 10;363(6429):543–6.

110. Kobayashi S, Yoshizawa H, Yamada S. Pathology of lumbar nerve root compression

Part 2: Morphological and immunohistochemical changes of dorsal root ganglion. J Orthop Res. 2004 Jan;22(1):180–8.

111. Aota Y, Onari K, An HS, Yoshikawa K. Dorsal Root Ganglia Morphologic Features in Patients With Herniation of the Nucleus Pulposus: Assessment Using Magnetic Resonance Myelography and Clinical Correlation. Spine. 2001 Oct 1;26(19):2125–32.

112. North RY, Li Y, Ray P, Rhines LD, Tatsui CE, Rao G, et al. Electrophysiological and transcriptomic correlates of neuropathic pain in human dorsal root ganglion neurons. Brain. 2019 May 1;142(5):1215–26.

113. Slipman CW, Isaac Z, Lenrow DA, Chou LH, Gilchrist RV, Vresilovic EJ. Clinical Evidence of Chemical Radiculopathy. 2002;5(3):6.

114. Kayama S, Konno S, Olmarker K, Yabuki S, Kikuchi S. Incision of the anulus fibrosus induces nerve root morphologic, vascular, and functional changes. An experimental study. Spine. 1996 Nov 15;21(22):2539–43.

115. Peng B, Wu W, Li Z, Guo J, Wang X. Chemical radiculitis: Pain. 2007 Jan;127(1):11–6.

116. Cyriax J. DURAL PAIN. The Lancet. 1978 Apr;311(8070):919–21.

117. Bove GM, Light AR. The nervi nervorum: Missing link for neuropathic pain? Pain Forum. 1997 Sep 1;6(3):181–90.

118. Hall TM, Elvey RL. Nerve trunk pain: physical diagnosis and treatment. Man Ther. 1999 May 1;4(2):63–73.

119. Marchettini P, Lacerenza M, Mauri E, Marangoni C. Painful Peripheral Neuropathies. Curr Neuropharmacol. 2006 Jul;4(3):175–81.

120. Meacham K, Shepherd A, Mohapatra DP, Haroutounian S. Neuropathic Pain: Central vs. Peripheral Mechanisms. Curr Pain Headache Rep. 2017 Jun;21(6):28

121. Haroutounian S, Nikolajsen L, Bendtsen TF, Finnerup NB, Kristensen AD, Hasselstrøm JB, et al. Primary afferent input critical for maintaining spontaneous pain in peripheral neuropathy. Pain. 2014 Jul;155(7):1272–9.

122. Rothman SM, Winkelstein BA. Chemical and mechanical nerve root insults induce differential behavioral sensitivity and glial activation that are enhanced in combination. Brain Res. 2007 Nov;1181:30–43.

123. Rutkowski MD, Winkelstein BA, Hickey WF, Pahl JL, DeLeo JA. Lumbar Nerve Root Injury Induces Central Nervous System Neuroimmune Activation and Neuroinflammation in the Rat: Relationship to Painful Radiculopathy. Spine. 2002 Aug;27(15):1604–13.

124. Schmidt CK, Rustagi T, Alonso F, Loukas M, Chapman JR, Oskouian RJ, et al. Nerve root anomalies: making sense of a complicated literature. Childs Nerv Syst. 2017 Aug;33(8):1261–73.

125. Latremoliere A, Woolf CJ. Central Sensitization: A Generator of Pain Hypersensitivity by Central Neural Plasticity. J Pain. 2009 Sep;10(9):895–926.

126. Calvo M, Dawes JM, Bennett DL. The role of the immune system in the generation of neuropathic pain. Lancet Neurol. 2012 Jul;11(7):629–42.

127. Albrecht DS, Ahmed SU, Kettner NW, Borra RJH, Cohen-Adad J, Deng H, et al.

Neuroinflammation of the spinal cord and nerve roots in chronic radicular pain patients: Pain. 2018 May;159(5):968–77.

128. Alshelh Z, Brusaferri L, Saha A, Morrissey E, Knight P, Kim M, et al. Neuroimmune signatures in chronic low back pain subtypes. Brain J Neurol. 2022 Apr 29;145(3):1098–110.

129. Frank MG, Fonken LK, Watkins LR, Maier SF. Microglia: Neuroimmune-sensors of stress. Semin Cell Dev Biol. 2019 Oct;94:176–85.

130. Beggs S, Liu XJ, Kwan C, Salter MW. Peripheral nerve injury and TRPV1-expressing primary afferent C-fibers cause opening of the blood-brain barrier. Mol Pain. 2010 Nov 2;6:74.

131. Ji RR, Berta T, Nedergaard M. Glia and pain: Is chronic pain a gliopathy?: Pain. 2013 Dec;154:S10–28.

132. Hunt JL, Winkelstein BA, Rutkowski MD, Weinstein JN, DeLeo JA. Repeated Injury to the Lumbar Nerve Roots Produces Enhanced Mechanical Allodynia and Persistent Spinal Neuroinflammation: Spine. 2001 Oct;26(19):2073–9.

133. Rosen S, Ham B, Mogil JS. Sex differences in neuroimmunity and pain. J Neurosci Res. 2017 Jan 2;95(1–2):500–8.

134. Smith BH, Torrance N. Epidemiology of Neuropathic Pain and Its Impact on Quality of Life. Curr Pain Headache Rep. 2012 Jun 1;16(3):191–8.

135. Konstantinou K, Dunn KM, Ogollah R, Vogel S, Hay EM. Characteristics of patients with low back and leg pain seeking treatment in primary care: baseline results from the ATLAS cohort study. BMC Musculoskelet Disord. 2015 Nov;16:332.

136. Pane K, Boccella S, Guida F, Franzese M, Maione S, Salvatore M. Role of gut microbiota in neuropathy and neuropathic pain states: A systematic preclinical review. Neurobiol Dis. 2022 Aug;170:105773.

137. Guo R, Chen LH, Xing C, Liu T. Pain regulation by gut microbiota: molecular mechanisms and therapeutic potential. Br J Anaesth. 2019 Nov;123(5):637–54.

138. Santoni M, Miccini F, Battelli N. Gut microbiota, immunity and pain. Immunol Lett. 2021 Jan;229:44–7.

139. Yang C, Fang X, Zhan G, Huang N, Li S, Bi J, et al. Key role of gut microbiota in anhedonia-like phenotype in rodents with neuropathic pain. Transl Psychiatry. 2019 Jan 31;9(1):1–11.

140. Bullmore ET. The inflamed mind: a radical new approach to depression. First U.S. edition. New York: Picador; 2019. 240 p.

141. Smith BH, Hébert HL, Veluchamy A. Neuropathic pain in the community: prevalence, impact, and risk factors. Pain. 2020 Sep;161 Suppl 1:S127–37.

142. Ford JJ, Kaddour O, Gonzales M, Page P, Hahne AJ. Clinical features as predictors of histologically confirmed inflammation in patients with lumbar disc herniation with associated radiculopathy. BMC Musculoskelet Disord. 2020 Aug 21;21(1):567.

143. Hider SL, Konstantinou K, Hay EM, Glossop J, Mattey DL. Inflammatory biomarkers do not distinguish between patients with sciatica and referred leg pain

within a primary care population: results from a nested study within the ATLAS cohort. BMC Musculoskelet Disord. 2019 Dec;20(1).

144. Jungen MJ, ter Meulen BC, van Osch T, Weinstein HC, Ostelo RWJG. Inflammatory biomarkers in patients with sciatica: a systematic review. BMC Musculoskelet Disord. 2019 Apr 9;20:156.

145. King AB. 'PHANTOM' SCIATICA. Arch Neurol Psychiatry. 1956 Jul 1;76(1):72.

146. Croci D, Fandino J, Marbacher S. Phantom Radiculopathy: Case Report and Review of the Literature. World Neurosurg. 2016 Jun;90:699.e19-699.e23.

147. Aydin SM, Zou SP, Varlotta G, Gharibo C. Successful Treatment of Phantom Radiculopathy with Fluoroscopic Epidural Steroid Injections. Pain Med. 2005 May 1;6(3):266-8.

148. Pham K, Gupta R. Understanding the mechanisms of entrapment neuropathies: Review article. Neurosurg Focus. 2009 Feb;26(2):E7.

149. Baron R, Maier C, Attal N, Binder A, Bouhassira D, Cruccu G, et al. Peripheral neuropathic pain: a mechanism-related organizing principle based on sensory profiles. PAIN. 2017 Feb;158(2):261-72.

150. International Association for the Study of Pain. Pain Terms and Definitions [Internet]. [cited 2023 Jan 20]. Available from: https://www.iasp-pain.org/resources/terminology/

151. Kosek E, Cohen M, Baron R, Gebhart GF, Mico JA, Rice ASC, et al. Do we need a third mechanistic descriptor for chronic pain states?: PAIN. 2016 Jul;157(7):1382-6.

152. Bennett MI, Attal N, Backonja MM, Baron R, Bouhassira D, Freynhagen R, et al. Using screening tools to identify neuropathic pain. PAIN. 2007 Feb;127(3):199-203.

153. Finnerup NB, Haroutounian S, Kamerman P, Baron R, Bennett DLH, Bouhassira D, et al. Neuropathic pain: an updated grading system for research and clinical practice. PAIN. 2016 Aug;157(8):1599-606.

154. Cruccu G, Truini A. A review of Neuropathic Pain: From Guidelines to Clinical Practice. Pain Ther. 2017 Dec;6(S1):35-42.

155. Ong BN, Konstantinou K, Corbett M, Hay E. Patients' Own Accounts of Sciatica: A Qualitative Study. Spine. 2011 Jul;36(15):1251-6.

156. Ryan C, Roberts L. 'Life on hold': The lived experience of radicular symptoms. A qualitative, interpretative inquiry. Musculoskelet Sci Pract. 2019 Feb;39:51-7.

157. Grøvle L, Haugen AJ, Natvig B, Brox JI, Grotle M. The prognosis of self-reported paresthesia and weakness in disc-related sciatica. Eur Spine J. 2013 Nov;22(11):2488-95.

158. Al Luwimi I, Ammar A, Al Awami M. Pathophysiology of Paresthesia. In: Imbelloni LE, editor. Paresthesia. InTech; 2012.

159. Filingeri D, Fournet D, Hodder S, Havenith G. Why wet feels wet? A neurophysiological model of human cutaneous wetness sensitivity. J Neurophysiol. 2014 Sep 15;112(6):1457-69.

160. Bannister K, Sachau J, Baron R, Dickenson AH. Neuropathic Pain: Mechanism-Based Therapeutics. Annu Rev Pharmacol Toxicol. 2020;60(1):257–74.
161. Viana F, Voets T. Heat Pain and Cold Pain. In: Wood JN, editor. The Oxford Handbook of the Neurobiology of Pain. Oxford University Press; 2020. p. 178–99.
162. Serra J, Solà R, Quiles C, Casanova-Molla J, Pascual V, Bostock H, et al. C-nociceptors sensitized to cold in a patient with small-fiber neuropathy and cold allodynia. Pain. 2009 Dec;147(1):46–53.
163. Yamaki S, Chau A, Gonzales L, McKemy DD. Nociceptive afferent phenotyping reveals that transient receptor potential ankyrin 1 promotes cold pain through neurogenic inflammation upstream of the neurotrophic factor receptor GFRα3 and the menthol receptor transient receptor potential melastatin 8. Pain. 2021 Feb;162(2):609–18.
164. Hakelius A, Nilsonne U, Pernow B, Zetterquist S. The Cold Sciatic Leg. Acta Orthop Scand. 1969 Jan;40(5):614–23.
165. Park TY, Son S, Lim TG, Jeong T. Hyperthermia associated with spinal radiculopathy as determined by digital infrared thermographic imaging. Medicine (Baltimore). 2020 Mar;99(11):e19483.
166. Segal JP, Tresidder KA, Bhatt C, Gilron I, Ghasemlou N. Circadian control of pain and neuroinflammation. J Neurosci Res. 2018 Jun;96(6):1002–20.
167. Nordin M, Nyström B, Wallin U, Hagbarth KE. Ectopic sensory discharges and paresthesiae in patients with disorders of peripheral nerves, dorsal roots and dorsal columns. Pain. 1984 Nov;20(3):231–45.
168. Defrin R, Brill S, Goor-Arieh I, Wood I, Devor M. 'Shooting pain' in lumbar radiculopathy and trigeminal neuralgia and ideas concerning its neural substrates. Pain. 2020;161(2):308-318.
169. Attal N, Bouhassira D. Mechanisms of pain in peripheral neuropathy. Acta Neurol Scand. 1999;100(s173):12–24.
170. Takahashi N, Yabuki S, Aoki Y, Kikuchi S. Pathomechanisms of Nerve Root Injury Caused by Disc Herniation: An Experimental Study of Mechanical Compression and Chemical Irritation. Spine. 2003 Mar;28(5):435–41.
171. Wu YS, Lin Y, Zhang XL, Tian NF, Sun LJ, Xu HZ, et al. The Projection of Nerve Roots on the Posterior Aspect of Spine From T11 to L5: A Cadaver and Radiological Study. Spine. 2012 Sep;37(20):E1232–7.
172. Crock HV. Normal and pathological anatomy of the lumbar spinal nerve root canals. J Bone Joint Surg Br. 1981;63(4):487–90.
173. Grimes PF, Massie JB, Garfin SR. Anatomic and Biomechanical Analysis of the Lower Lumbar Foraminal Ligaments: Spine. 2000 Aug;25(16):2009–14.
174. Spencer DL, Irwin GS, Miller JA. Anatomy and significance of fixation of the lumbosacral nerve roots in sciatica. Spine. 1983 Sep;8(6):672–9.
175. Kraan GA. The extraforaminal ligaments of the human spinal nerves: anatomical and biomechanical study. 2021. [PhD-Thesis - Research and graduation internal, Vrije Universiteit Amsterdam].
176. Suh SW, Shingade VU, Lee SH, Bae JH, Park CE, Song JY. Origin of lumbar spinal

roots and their relationship to intervertebral discs: a cadaver and radiological study. J Bone Joint Surg Br. 2005 Apr;87-B(4):518–22.

177. Parisien RC, Ball PA. William Jason Mixter (1880-1958). Ushering in the 'dynasty of the disc.' Spine. 1998 Nov 1;23(21):2363–6.

178. Goldthwait JE. The Lumbo-Sacral Articulation; An Explanation of Many Cases of 'Lumbago,' 'Sciatica' and Paraplegia. Boston Med Surg J. 1911 Mar 16;164(11):365–72.

179. Middleton GS, Teacher JH. Injury of the Spinal Cord Due to Rupture of an Intervertebral Disc during Muscular Effort 1Read at a meeting of the Medico-Chirurgical Society of Glasgow held on 3rd March, 1911. Glasg Med J. 1911 Jul;76(1):1–6.

180. Dandy WE. Loose Cartilage From Intervertebral Disk Simulating Tumor of the Spinal Cord. Arch Surg. 1929 Oct 1;19(4):660–72.

181. Blamoutier A. Nerve root compression by lumbar disc herniation: A french discovery? Orthop Traumatol Surg Res. 2019 Apr 1;105(2):335–8.

182. Mixter WJ, Barr JS. Rupture of the Intervertebral Disc with Involvement of the Spinal Canal. N Engl J Med. 1934 Aug 2;211(5):210–5.

183. Allan DB, Waddell G. An historical perspective on low back pain and disability. Acta Orthop Scand. 1989 Jan;60(sup234):1–23.

184. Saker E, Tubbs S. Anatomy of the Lumbar Intervertebral Discs [Internet]. 2018 [cited 2019 Aug 18]. Available from: http://www.wheelessonline.com/ISSLS/lumbar-intervertebral-discs-anatomy/

185. Pinheiro-Franco JL, Vaccaro AR, Benzel EC, Mayer HM, editors. Advanced Concepts in Lumbar Degenerative Disk Disease. Berlin, Heidelberg: Springer Berlin Heidelberg; 2016.

186. Bogduk N. Clinical and radiological anatomy of the lumbar spine. 5th ed. Edinburgh; New York: Churchill Livingstone; 2012.

187. Adams MA, Dolan P. Intervertebral disc degeneration: evidence for two distinct phenotypes. J Anat. 2012;221(6):497–506.

188. Veres SP, Robertson PA, Broom ND. ISSLS Prize Winner: Microstructure and Mechanical Disruption of the Lumbar Disc Annulus: Part II: How the Annulus Fails Under Hydrostatic Pressure. Spine. 2008 Dec;33(25):2711–20.

189. Wade K, Berger-Roscher N, Saggese T, Rasche V, Wilke H. How annulus defects can act as initiation sites for herniation. Eur Spine J. 2022 Jun;31(6):1487–500.

190. Suri P, Hunter DJ, Jouve C, Hartigan C, Limke J, Pena E, et al. Inciting Events Associated with Lumbar Disk Herniation. Spine J Off J North Am Spine Soc. 2010 May;10(5):388–95.

191. Alyas F, Connell D, Saifuddin A. Upright positional MRI of the lumbar spine. Clin Radiol. 2008 Sep;63(9):1035–48.

192. Jinkins JR, Dworkin JS, Green CA, Greenhalgh JF, Gianni M, Gelbien M, et al. Upright, Weight-Bearing, Dynamic-Kinetic MRI of the Spine. Riv Neuroradiol. 2002;25.

193. Shapiro IM, Risbud MV, editors. The Intervertebral Disc: Molecular and Structural Studies of the Disc in Health and Disease. Vienna: Springer Vienna; 2014.

194. Singh H, Moss I. The Intervertebral Disc: Function and Degradation [Internet]. 2018 [cited 2019 Aug 18]. Available from: http://www.wheelessonline.com/ISSLS/ section-2-chapter-2-the-intervertebral-disc-function-and-degradation/

195. Cook JL, Purdam CR. Is tendon pathology a continuum? A pathology model to explain the clinical presentation of load-induced tendinopathy. Br J Sports Med. 2009 Jun 1;43(6):409–16.

196. Skrzypiec D, Tarala M, Pollintine P, Dolan P, Adams MA. When are intervertebral discs stronger than their adjacent vertebrae? Spine. 2007 Oct 15;32(22):2455–61.

197. Videman T, Gibbons LE, Kaprio J, Battié MC. Outstanding Paper: Medical and InterventionalScience Challenging the Cumulative Injury Model: Positive Effects of Greater Body Mass on Disc Degeneration. Spine J. 2009 Oct;9(10):40S.

198. Maurer E, Klinger C, Lorbeer R, Rathmann W, Peters A, Schlett CL, et al. Long-term effect of physical inactivity on thoracic and lumbar disc degeneration—an MRI-based analysis of 385 individuals from the general population. Spine J. 2020 Sep 1;20(9):1386–96.

199. Belavý DL, Quittner MJ, Ridgers N, Ling Y, Connell D, Rantalainen T. Running exercise strengthens the intervertebral disc. Sci Rep. 2017 Apr 19;7(1):45975.

200. Owen PJ, Hangai M, Kaneoka K, Rantalainen T, Belavy DL. Mechanical loading influences the lumbar intervertebral disc. A cross-sectional study in 308 athletes and 71 controls. J Orthop Res. 2021 May;39(5):989-997.

201. Teichtahl AJ, Urquhart DM, Wang Y, Wluka AE, O'Sullivan R, Jones G, et al. Physical inactivity is associated with narrower lumbar intervertebral discs, high fat content of paraspinal muscles and low back pain and disability. Arthritis Res Ther. 2015 May 7;17:114.

202. Belavy DL, Adams M, Brisby H, Cagnie B, Danneels L, Fairbank J, et al. Disc herniations in astronauts: What causes them, and what does it tell us about herniation on earth? Eur Spine J. 2016 Jan 1;25(1):144–54.

203. Zhang Y, Sun Z, Liu J, Guo X. Advances in Susceptibility Genetics of Intervertebral Degenerative Disc Disease. Int J Biol Sci. 2008 Sep 2;4(5):283–90.

204. Patel AA, Spiker WR, Daubs M, Brodke D, Cannon-Albright LA. Evidence for an Inherited Predisposition to Lumbar Disc Disease. J Bone Joint Surg Am. 2011 Feb 2;93(3):225–9.

205. Battié MC, Videman T, Kaprio J, Gibbons LE, Gill K, Manninen H, et al. The Twin Spine Study: Contributions to a changing view of disc degeneration. Spine J. 2009 Jan;9(1):47–59.

206. Sambrook PN, MacGregor AJ, Spector TD. Genetic influences on cervical and lumbar disc degeneration: A magnetic resonance imaging study in twins. Arthritis Rheum. 1999;42(2):366–72.

207. Walls CB, Snell A, McLean DJ, Pearce N. An updated critique of the use of the Twin Spine Study (2009) to determine causation of low back disorder. N Z Med J. 2019 May 3;132(1494):57–9.

208. Eskola PJ, Lemmelä S, Kjaer P, Solovieva S, Männikkö M, Tommerup N, et al.

Genetic Association Studies in Lumbar Disc Degeneration: A Systematic Review. PLOS ONE. 2012 Nov 21;7(11):e49995.

209. Adams MA, Lama P, Zehra U, Dolan P. Why do some intervertebral discs degenerate, when others (in the same spine) do not?: Why Do Some Intervertebral Discs Degenerate? Clin Anat. 2015 Mar;28(2):195–204.

210. Jiang L, Yuan F, Yin X, Dong J. Responses and adaptations of intervertebral disc cells to microenvironmental stress: a possible central role of autophagy in the adaptive mechanism. Connect Tissue Res. 2014 Dec;55(5–6):311–21.

211. Molinos M, Almeida CR, Caldeira J, Cunha C, Gonçalves RM, Barbosa MA. Inflammation in intervertebral disc degeneration and regeneration. J R Soc Interface. 2015 Mar 6;12(104):20141191.

212. Nerlich AG, Schaaf R, Wälchli B, Boos N. Temporo-spatial distribution of blood vessels in human lumbar intervertebral discs. Eur Spine J. 2007 Apr;16(4):547–55.

213. Adams MA, Stefanakis M, Dolan P. Healing of a painful intervertebral disc should not be confused with reversing disc degeneration: Implications for physical therapies for discogenic back pain. Clin Biomech. 2010 Dec 1;25(10):961–71.

214. Tawa N, Rhoda A, Diener I. Accuracy of magnetic resonance imaging in detecting lumbo-sacral nerve root compromise: a systematic literature review. BMC Musculoskelet Disord. 2016; 17(1): 386.

215. Nordberg CL, Boesen M, Fournier GL, Bliddal H, Hansen P, Hansen BB. Positional changes in lumbar disc herniation during standing or lumbar extension: a cross-sectional weight-bearing MRI study. Eur Radiol. 2021;31:804–812

216. Seo JY, Roh YH, Kim YH, Ha KY. Three-dimensional analysis of volumetric changes in herniated discs of the lumbar spine: does spontaneous resorption of herniated discs always occur? Eur Spine J. 2016 May;25(5):1393–402.

217. Chiu CC, Chuang TY, Chang KH, Wu CH, Lin PW, Hsu WY. The probability of spontaneous regression of lumbar herniated disc: a systematic review. Clin Rehabil. 2015 Feb;29(2):184–95.

218. Ido K, Urushidani H. Fibrous adhesive entrapment of lumbosacral nerve roots as a cause of sciatica. Spinal Cord. 2001 May;39(5):269–73.

219. Porter RW, Hibbert C, Evans C. The natural history of root entrapment syndrome. Spine. 1984 Jun;9(4):418–21.

220. O'Neill CW, Kurgansky ME, Derby R, Ryan DP. Disc Stimulation and Patterns of Referred Pain: Spine. 2002 Dec;27(24):2776–81.

221. Buijs E, Visser L, Groen G. Sciatica and the sacroiliac joint: a forgotten concept. BJA Br J Anaesth. 2007 Nov 1;99(5):713–6.

222. Hopayian K, Heathcote J. Deep gluteal syndrome: an overlooked cause of sciatica. Br J Gen Pract. 2019 Aug 28;69(687):485–6.

223. Gandhi J, Wilson AL, Liang R, Weissbart SJ, Khan SA. Sciatic endometriosis: A narrative review of an unusual neurogynecologic condition. J Endometr Pelvic Pain Disord. 2021 Mar 1;13(1):3–9.

224. Moore RJ, Vernon-Roberts B, Fraser RD, Osti OL, Schembri M. The Origin and

Fate of Herniated Lumbar Intervertebral Disc Tissue. Spine. 1996 Sep 15;21(18):2149–55.

225. Willburger RE, Ehiosun UK, Kuhnen C, Krämer J, Schmid G. Clinical Symptoms in Lumbar Disc Herniations and Their Correlation to the Histological Composition of the Extruded Disc Material: Spine. 2004 Aug;29(15):1655–61.

226. Lama P, Zehra U, Balkovec C, Claireaux HA, Flower L, Harding IJ, et al. Significance of cartilage endplate within herniated disc tissue. Eur Spine J. 2014 Sep;23(9):1869–77.

227. Brock M, Patt S, Mayer HM. The form and structure of the extruded disc. Spine. 1992 Dec;17(12):1457–61.

228. Yasuma T, Arai K, Yamauchi Y. The histology of lumbar intervertebral disc herniation. The significance of small blood vessels in the extruded tissue. Spine. 1993 Oct 1;18(13):1761–5.

229. Rajasekaran S, Bajaj N, Tubaki V, Kanna RM, Shetty AP. ISSLS Prize winner: The anatomy of failure in lumbar disc herniation: an in vivo, multimodal, prospective study of 181 subjects. Spine. 2013 Aug 1;38(17):1491–500.

230. Sahoo MM, Mahapatra SK, Kaur S, Sarangi J, Mohapatra M. Significance of Vertebral Endplate Failure in Symptomatic Lumbar Disc Herniation. Glob Spine J. 2017 May;7(3):230–8.

231. Goldie I. Granulation Tissue in the Ruptured Intervertebral Disc. Acta Pathol Microbiol Scand. 1958;42(4):302–4.

232. Kokubo Y, Uchida K, Kobayashi S, Yayama T, Sato R, Nakajima H, et al. Herniated and spondylotic intervertebral discs of the human cervical spine: histological and immunohistological findings in 500 en bloc surgical samples: Laboratory investigation. J Neurosurg Spine. 2008 Sep;9(3):285–95.

233. Fardon DF, Williams AL, Dohring EJ, Murtagh FR, Gabriel Rothman SL, Sze GK. Lumbar disc nomenclature: version 2.0. Spine J. 2014 Nov;14(11):2525–45.

234. Khan JM, McKinney D, Basques BA, Louie PK, Carroll D, Paul J, et al. Clinical Presentation and Outcomes of Patients With a Lumbar Far Lateral Herniated Nucleus Pulposus as Compared to Those With a Central or Paracentral Herniation. Glob Spine J. 2019 Aug 1;9(5):480–6.

235. Vroomen P, Wilmink J, de Krom M. Prognostic value of MRI findings in sciatica. Neuroradiology. 2002 Jan;44(1):59–63.

236. Brinjikji W, Luetmer PH, Comstock B, Bresnahan BW, Chen LE, Deyo RA, et al. Systematic Literature Review of Imaging Features of Spinal Degeneration in Asymptomatic Populations. Am J Neuroradiol. 2015 Apr;36(4):811–6.

237. Brinjikji W, Diehn FE, Jarvik JG, Carr CM, Kallmes DF, Murad MH, et al. MRI Findings of Disc Degeneration are More Prevalent in Adults with Low Back Pain than in Asymptomatic Controls: A Systematic Review and Meta-Analysis. Am J Neuroradiol. 2015 Dec 1;36(12):2394–9.

238. Suri P, Boyko EJ, Goldberg J, Forsberg CW, Jarvik JG. Longitudinal associations between incident lumbar spine MRI findings and chronic low back pain or radicular symptoms: retrospective analysis of data from the longitudinal assessment of

imaging and disability of the back (LAIDBACK). BMC Musculoskelet Disord. 2014;15:152.

239. Boos N, Rieder R, Schade V, Spratt KF, Semmer N, Aebi M. The Diagnostic Accuracy of Magnetic Resonance Imaging, Work Perception, and Psychosocial Factors in Identifying Symptomatic Disc Herniations. Spine. 1995 Dec 15;20(24):2613–25.

240. van Rijn JC, Klemetso N, Reitsma JB, Majoie CBLM, Hulsmans FJ, Peul WC, et al. Symptomatic and asymptomatic abnormalities in patients with lumbosacral radicular syndrome: Clinical examination compared with MRI. Clin Neurol Neurosurg. 2006 Sep;108(6):553–7.

241. Barth M, Diepers M, Weiss C, Thomé C. Two-year outcome after lumbar microdiscectomy versus microscopic sequestrectomy: part 2: radiographic evaluation and correlation with clinical outcome. Spine. 2008 Feb 1;33(3):273–9.

242. Fraser R, Sandhu A, Gogan W. Magnetic resonance imaging findings 10 years after treatment for lumbar disc herniation - PubMed. Spine. 1995;20(6):710–4.

243. Pople IK, Griffith HB. Prediction of an extruded fragment in lumbar disc patients from clinical presentations. Spine. 1994 Jan 15;19(2):156–8.

244. Reihani-Kermani H. Clinical aspects of sciatica and their relation to the type of lumbar disc herniation. Arch of Iran Med. 2005;8.

245. Dunsmuir RA, Nisar S, Cruickshank JA, Loughenbury PR. No correlation identified between the proportional size of a prolapsed intravertebral disc with disability or leg pain. Bone Jt J. 2022 Jun;104-B(6):715–20.

246. Karppinen J, Malmivaara A, Tervonen O, Pääkkö E, Kurunlahti M, Syrjälä P, et al. Severity of symptoms and signs in relation to magnetic resonance imaging findings among sciatic patients. Spine. 2001 Apr 1;26(7):E149-154.

247. Mariajoseph FP, Castle-Kirszbaum M, Kam J, Rogers M, Sher R, Daly C, et al. Relationship between herniated intervertebral disc fragment weight and pain in lumbar microdiscectomy patients. J Clin Neurosci Off J Neurosurg Soc Australas. 2022 Aug;102:75–9.

248. Nv A, Rajasekaran S, Ks SVA, Kanna RM, Shetty AP. Factors that influence neurological deficit and recovery in lumbar disc prolapse—a narrative review. Int Orthop. 2019 Apr;43(4):947–55.

249. Gupta A, Upadhyaya S, Yeung CM, Ostergaard PJ, Fogel HA, Cha T, et al. Does Size Matter? An Analysis of the Effect of Lumbar Disc Herniation Size on the Success of Nonoperative Treatment. Glob Spine J. 2022;10(7):881–887.

250. Masui T, Yukawa Y, Nakamura S, Kajino G, Matsubara Y, Kato F, et al. Natural History of Patients with Lumbar Disc Herniation Observed by Magnetic Resonance Imaging for Minimum 7 Years: J Spinal Disord Tech. 2005 Apr;18(2):121–6.

251. Modic MT, Obuchowski NA, Ross JS, Brant-Zawadzki MN, Grooff PN, Mazanec DJ, et al. Acute Low Back Pain and Radiculopathy: MR Imaging Findings and Their Prognostic Role and Effect on Outcome. Radiology. 2005 Nov;237(2):597–604.

252. Kim JH, van Rijn RM, van Tulder MW, Koes BW, de Boer MR, Ginai AZ, et al. Diagnostic accuracy of diagnostic imaging for lumbar disc herniation in adults with

low back pain or sciatica is unknown; a systematic review. Chiropr Man Ther. 2018 Aug;26:37

253. Weiner BK, Patel R. The accuracy of MRI in the detection of Lumbar Disc Containment. J Orthop Surg. 2008 Oct 2;3:46.

254. Ahn SH, Ahn MW, Byun WM. Effect of the Transligamentous Extension of Lumbar Disc Herniations on Their Regression and the Clinical Outcome of Sciatica: Spine. 2000 Feb;25(4):475–80.

255. Fagerlund MK, Thelander U, Friberg S. Size of lumbar disc hernias measured using computed tomography and related to sciatic symptoms. Acta Radiol Stockh Swed 1987. 1990 Nov;31(6):555–8.

256. Kesikburun B, Eksioglu E, Turan A, Adiguzel E, Kesikburun S, Cakci A. Spontaneous regression of extruded lumbar disc herniation: Correlation with clinical outcome. Pak J Med Sci. 2019;35(4):974–80.

257. Komori H, Shinomiya K, Nakai O, Yamaura I, Takeda S, Furuya K. The natural history of herniated nucleus pulposus with radiculopathy. Spine. 1996 Jan 15;21(2):225–9.

258. Phillips K, Chiverton N, Michael A, Cole A, Breakwell L, G H, et al. The cytokine and chemokine expression profile of nucleus pulposus cells: implications for degeneration and regeneration of the intervertebral disc. Arthritis Res Ther. 2013 Jan 1;15(6):R213–R213.

259. Jones P, Gardner L, Menage J, Williams GT, Roberts S. Intervertebral disc cells as competent phagocytes in vitro: implications for cell death in disc degeneration. Arthritis Res Ther. 2008;10(4):R86.

260. Saal JS, Franson RC, Dobrow R, Saal JA, White AH, Goldthwaite N. High levels of inflammatory phospholipase A2 activity in lumbar disc herniations. Spine. 1990 Jul;15(7):674–8.

261. Jonsson D, Olmarker K. Pathophysiology of Lumbar Radiculopathy [Internet]. 2018 [cited 2019 Aug 18]. Available from: http://www.wheelessonline.com/ISSLS/section-2-chapter-8-pathophysiology-of-lumbar-radiculopathy/

262. Byröd G, Otani K, Brisby H, Rydevik B, Olmarker K. Methylprednisolone reduces the early vascular permeability increase in spinal nerve roots induced by epidural nucleus pulposus application. J Orthop Res Off Publ Orthop Res Soc. 2000 Nov;18(6):983–7.

263. Andrade P, van Aalst J, Bauwens M, Vogg A, van Kroonenburgh MJ, Mottaghy FM, et al. Radionuclide tumor necrosis factor-alpha activity in herniated lumbar disc correlates with severe leg pain. Surg Neurol Int. 2020;11:344

264. Wang YF, Chen PY, Chang W, Zhu FQ, Xu LL, Wang SL, et al. Clinical significance of tumor necrosis factor-α inhibitors in the treatment of sciatica: a systematic review and meta-analysis. PloS One. 2014;9(7):e103147.

265. Cunha C, Silva AJ, Pereira P, Vaz R, Gonçalves RM, Barbosa MA. The inflammatory response in the regression of lumbar disc herniation. Arthritis Res Ther. 2018; 20: 251

266. Di Martino A, Merlini L, Faldini C. Autoimmunity in intervertebral disc herniation: from bench to bedside. Expert Opin Ther Targets. 2013 Dec;17(12):1461–70.
267. Sun Z, Liu B, Luo ZJ. The Immune Privilege of the Intervertebral Disc: Implications for Intervertebral Disc Degeneration Treatment. Int J Med Sci. 2020;17(5):685–92.
268. Bermudez-Lekerika P, Crump KB, Tseranidou S, Nüesch A, Kanelis E, Alminnawi A, et al. Immuno-Modulatory Effects of Intervertebral Disc Cells. Front Cell Dev Biol. 2022 Jun 29;10:924692.
269. Bobechko WP, Hirsch C. Auto-immune Response to Nucleus Puloposus in the Rabbit. J Bone Joint Surg Br. 1965 Aug;47:574–80.
270. Marshall LL, Trethewie ER, Curtain CC. Chemical radiculitis. A clinical, physiological and immunological study. Clin Orthop. 1977 Dec;(129):61–7.
271. Jin L, Xiao L, Ding M, Pan A, Balian G, Sung SSJ, et al. Heterogeneous macrophages contribute to the pathology of disc herniation induced radiculopathy. Spine J. 2022 Apr;22(4):677–89.
272. Djuric N, Yang X, El Barzouhi A, Ostelo R, van Duinen SG, Lycklama À Nijeholt GJ, et al. Lumbar disc extrusions reduce faster than bulging discs due to an active role of macrophages in sciatica. Acta Neurochir. 2020;162(1):79-85.
273. Kobayashi S, Meir A, Kokubo Y, Uchida K, Takeno K, Miyazaki T, et al. Ultrastructural Analysis on Lumbar Disc Herniation Using Surgical Specimens: Role of Neovascularization and Macrophages in Hernias. Spine. 2009 Apr;34(7):655–62.
274. Cribb G, Jaffray D, Cassar-Pullicino V. Observations on the natural history of massive lumbar disc herniation. J Bone Joint Surg Br. 2007 Jul 1;89:782–4.
275. Ikeda T, Nakamura T, Kikuchi T, Umeda S, Senda H, Takagi K. Pathomechanism of spontaneous regression of the herniated lumbar disc: histologic and immunohistochemical study. J Spinal Disord. 1996 Apr;9(2):136–40.
276. Wang Y, Dai G, Jiang L, Liao S. The incidence of regression after the non-surgical treatment of symptomatic lumbar disc herniation: a systematic review and meta-analysis. BMC Musculoskelet Disord. 2020 Aug 10;21(1):530.
277. Adams MA, Hutton WC. Gradual disc prolapse. Spine. 1985 Aug;10(6):524–31.
278. Macnab I. Negative disc exploration. An analysis of the causes of nerve-root involvement in sixty-eight patients. J Bone Joint Surg Am. 1971 Jul;53(5):891–903.
279. Burton CV, Kirkaldy-Willis WH, Yong-Hing K, Heithoff KB. Causes of failure of surgery on the lumbar spine. Clin Orthop. 1981 Jun;(157):191–9.
280. Yuan S guo, Wen Y liang, Zhang P, Li Y kai. Ligament, nerve, and blood vessel anatomy of the lateral zone of the lumbar intervertebral foramina. Int Orthop. 2015 Nov;39(11):2135–41.
281. Orita S, Inage K, Eguchi Y, Kubota G, Aoki Y, Nakamura J, et al. Lumbar foraminal stenosis, the hidden stenosis including at L5/S1. Eur J Orthop Surg Traumatol. 2016 Oct 1;26(7):685–93.
282. Hasegawa T, An HS, Haughton VM, Nowicki BH. Lumbar foraminal stenosis: critical heights of the intervertebral discs and foramina. A cryomicrotome study in cadavera. J Bone Joint Surg Am. 1995 Jan;77(1):32–8.

283. Al-Rawahi M, Luo J, Pollintine P, Dolan P, Adams MA. Mechanical function of vertebral body osteophytes, as revealed by experiments on cadaveric spines. Spine. 2011 May 1;36(10):770–7.

284. Jenis LG, An HS. Spine Update: Lumbar Foraminal Stenosis. Spine. 2000 Feb;25(3):389–94.

285. Yamada K, Aota Y, Higashi T, Ishida K, Nimura T, Konno T, et al. Lumbar foraminal stenosis causes leg pain at rest. Eur Spine J. 2014 Mar;23(3):504–7.

286. Mostofi K, Peyravi M, Moghaddam BG. A comparison of sciatica in young subjects and elderly person. J Clin Orthop Trauma. 2020 Oct;11(5):889–90.

287. Schmid AB, Tampin B, Baron R, Finnerup NB, Hansson P, Hietaharju A, et al. Recommendations for terminology and the identification of neuropathic pain in people with spine-related leg pain. Outcomes from the NeuPSIG working group. PAIN. 2022 May

'*Remember, always, that everything you know, and everything everyone knows, is only a model. Get your model out there where it can be viewed. Invite others to challenge your assumptions and add their own. Instead of becoming a champion for one possible explanation or hypothesis or model, collect as many as possible. Consider all of them to be plausible until you find some evidence that causes you to rule them out.*'

— DONELLA MEADOWS, 2008

ACKNOWLEDGMENTS

I'm fortunate to belong to a generous profession. My colleagues are eager to learn, eager to teach what they know and enthusiastic in their support for one another's endeavours. I'm grateful to everyone I have spoken to for their knowledge, guidance and encouragement. This book is very much the product of a community of practice.

In particular, I'd like to express my gratitude to everyone who kindly took the time to be interviewed on my podcast: David Butler, Kate Charlton (twice!), Michelle Angus, Drew Jordan, Giacomo Carta, Mark Laslett, Daniel Albrecht, Raymond Ostelo, Adam Dobson and Harris Ashraff.

I'm also grateful to David Poulter for continually pointing out new trails for me to explore, to Christine Price for helping me to see things I hadn't seen, to Jack Chew for having faith that I might have something useful to say.

Of course, I'm particularly grateful to Annina for first sparking my interest in nerve pain, for illuminating things that didn't make sense, and for taking the time to help me write this book. With her expertise and direction, she has raised the book far above its original form.

- Tom Jesson

ABOUT THE AUTHORS

Tom Jesson is a physiotherapist by trade. He has worked with people with a wide range of musculoskeletal conditions, including within in a specialist pain service. Tom has been writing for many years, but first started to focus his work on sciatica when he noticed that many other clinicians had the same burning questions he did - What is sciatica? How can we best make clinical decisions about it? And what on earth can we do to treat it?

Since focusing on sciatica, Tom has published peer reviewed work, two books (the one you are reading and one, with Rob Tyer, on cauda equina syndrome) and dozens to hundreds of comics, blogs, podcasts and newsletters. His work can be found on his Substack or his personal site.

Tom is originally from Northumberland, England, but he now lives in Houston, Texas with his wife and two daughters. When he is not writing about sciatica, he would like to say that he's reading a good book or going for a long run, but really he's probably changing a diaper or trying to get a baby to go to sleep.

Prof. Annina Schmid is a Specialist Musculoskeletal Physiotherapist and a Neuroscientist affiliated with the Nuffield Department of Clinical Neurosciences at Oxford University in the UK. She has a strong interest in entrapment neuropathies and neuropathic pain as well as the development of precision therapy for these patients.

After her clinical training in Switzerland and Australia, she realised that she only scratches the surface of pain and nerve injuries and decided to move to the dark side (translational research). Her work spans all the way from animal models of nerve injuries to growing human sensory neurones in a dish and prodding people to assess their pain clinically or image their nerves with advanced neuroimaging techniques. She leads the Neuromusculoskeletal Health and Science Lab at Oxford University and her team uses a translational and interdisciplinary approach to study the pathophysiology of neuromusculoskeletal conditions with the ultimate goal to improve management for patients.

Annina's research contributions have been recognised by the award of several prizes (most recently the Emerging Leaders Prize in Pain Research, Medical Research Foundation) and competitive fellowships. She was the first allied health professional to win a prestigious Clinical Research Career Development Fellowship from the Wellcome Trust, recognising her contributions to discovery sciences.

In addition to her research activities, Annina teaches postgraduate courses related to pain and neuroscience internationally. She also maintains a weekly clinic as a specialist musculoskeletal Physiotherapist both privately and in a public chronic pain programme. If she is not thinking about nerve fibres, immune cells or electrical potentials, she likes to immerse in Argentine Tango dancing.

Further information about Annina can be found at www.neuro-research.ch.

Peter Jesson is Tom's dad. A retired civil servant, he has been biding his time for decades, waiting for the opportunity to unleash his creative side and illustrate a book about sciatica. He is happy that the long hours he spent as a teenager replicating the covers of naff prog rock albums have finally borne fruit - it turns out he's a pretty decent artist.

ALSO BY TOM JESSON

Cauda Equina Syndrome: The MSK Clinician's Guide.

Practical, interesting and never, ever boring. A book to help you meet the challenge of potential CES in your practice.

'Extremely high value information. I feel like this is the type of stuff that separates expert clinicians from the others!'

— TIM, PHYSIO.

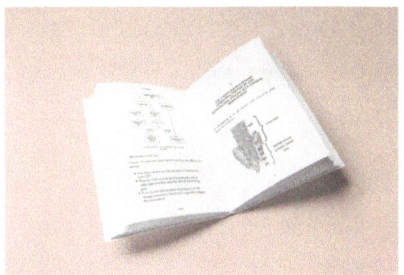